TRANSCENDENTAL NUMBER THEORY

First published in 1975, this classic book gives a systematic account of transcendental number theory, that is, the theory of those numbers that cannot be expressed as the roots of algebraic equations having rational coefficients. Their study has developed into a fertile and extensive theory, which continues to see rapid progress today. Expositions are presented of theories relating to linear forms in the logarithms of algebraic numbers, of Schmidt's generalization of the Thue–Siegel–Roth theorem, of Shidlovsky's work on Siegel's E-functions and of Sprindžuk's solution to the Mahler conjecture.

This edition includes an introduction written by David Masser describing Baker's achievement, surveying the content of each chapter and explaining the main argument of Baker's method in broad strokes. A new afterword lists recent developments related to Baker's work.

Alan Baker was one of the leading British mathematicians of the past century. He took great strides in number theory by, among other achievements, obtaining a vast generalization of the Gelfond–Schneider Theorem and using it to give effective solutions to a large class of Diophantine problems. This work kicked off a new era in transcendental number theory and won Baker the Fields Medal in 1970.

David Masser is Professor Emeritus in the Department of Mathematics and Computer Science at the University of Basel. He is a leading researcher in transcendence methods and applications and helped correct the proofs of the original edition of *Transcendental Number Theory* as Baker's student.

CAMBRIDGE
MATHEMATICAL LIBRARY

Cambridge University Press has a long and honourable history of publishing in mathematics and counts many classics of the mathematical literature within its list. Some of these titles have been out of print for many years now and yet the methods which they espouse are still of considerable relevance today.

The *Cambridge Mathematical Library* provides an inexpensive edition of these titles in a durable paperback format and at a price that will make the books attractive to individuals wishing to add them to their own personal libraries. Certain volumes in the series have a foreword, written by a leading expert in the subject, which places the title in its historical and mathematical context.

A complete list of books in the series can be found at www.cambridge.org/mathematics. Recent titles include the following:

Attractors for Semigroups and Evolution Equations
OLGA A. LADYZHENSKAYA

Fourier Analysis
T. W. KÖRNER

Transcendental Number Theory
ALAN BAKER

An Introduction to Symbolic Dynamics and Coding (Second Edition)
DOUGLAS LIND & BRIAN MARCUS

Reversibility and Stochastic Networks
F. P. KELLY

The Geometry of Moduli Spaces of Sheaves (Second Edition)
DANIEL HUYBRECHTS & MANFRED LEHN

Smooth Compactifications of Locally Symmetric Varieties (Second Edition)
AVNER ASH, DAVID MUMFORD, MICHAEL RAPOPORT &
YUNG-SHENG TAI

Markov Chains and Stochastic Stability (Second Edition)
SEAN MEYN & RICHARD L. TWEEDIE

TRANSCENDENTAL NUMBER THEORY

ALAN BAKER F.R.S.

With an Introduction by

DAVID MASSER
Universität Basel, Switzerland

CAMBRIDGE
UNIVERSITY PRESS

CAMBRIDGE
UNIVERSITY PRESS

University Printing House, Cambridge CB2 8BS, United Kingdom

One Liberty Plaza, 20th Floor, New York, NY 10006, USA

477 Williamstown Road, Port Melbourne, VIC 3207, Australia

314–321, 3rd Floor, Plot 3, Splendor Forum, Jasola District Centre, New Delhi – 110025, India

103 Penang Road, #05–06/07, Visioncrest Commercial, Singapore 238467

Cambridge University Press is part of the University of Cambridge.

It furthers the University's mission by disseminating knowledge in the pursuit of education, learning, and research at the highest international levels of excellence.

www.cambridge.org
Information on this title: www.cambridge.org/9781009229944
DOI: 10.1017/9781009229937

First published 1975
Reprinted with additional material 1979
Reissued as a paperback with updated material in the Cambridge Mathematical Library series 1990
Reprinted with introduction 2022

A catalogue record for this publication is available from the British Library.

ISBN 978-1-009-22994-4 Paperback

Contents

Introduction

David Masser

On the first page of the Bibliography are listed earlier works about some of the topics treated in this monograph. In particular the books of Gelfond, Schneider and Siegel are universally regarded as milestones in the development of the theory of transcendental numbers. Each book was based largely on the author's own breakthroughs.

The present monograph represented a similar milestone. Chapters 2, 3, 4, 5, 9, 10, and to a lesser extent Chapters 6, 8, cover material due to the author Alan Baker. This material and Baker's own further developments of it earned him a Fields Medal in 1970.

Of course it is the material in Chapter 2 that constitutes the heart of his achievement. This is explained in the first two pages with a characteristic brevity and modesty. Here we wish to complement this with the following less brief and modest account.

The essential ideas can be conveyed through the special case of his Theorem 2.1 for $n = 2, 3$, even ignoring the extra 1 that appears there.

We start with $n = 2$. It amounts to the impossibility of

$$\beta \log \alpha = \log \alpha' \tag{1}$$

for α, α' non-zero algebraic numbers and β irrational algebraic. Of course this is the Gelfond–Schneider Theorem of 1934. It also follows from Theorem 6.1 of Chapter 6, and we proceed to sketch the argument.

We assume (1) and we will obtain a contradiction. Following Gelfond we construct a non-zero polynomial F, say in $\mathbf{Z}[x, y]$, such that

$$f(z) = F(e^z, e^{\beta z}) \tag{2}$$

has many zeroes. More precisely we need the derivatives

$$f^{(t)}(s \log \alpha) = 0 \tag{3}$$

for a certain range of integers

$$t < T, \quad 1 \leq s \leq S \tag{4}$$

with T, S integers to be suitably chosen later. Thus we have zeroes at $\log\alpha, \ldots, S\log\alpha$ and moreover of multiplicity at least T. The point is that the functions $e^z, e^{\beta z}$ in (2) take the values

$$e^{\log\alpha} = \alpha, \quad e^{\beta\log\alpha} = e^{\log\alpha'} = \alpha'$$

at say $z = \log\alpha$; and these are algebraic numbers. Similarly at $s\log\alpha$ and with multiplicities. Thus the conditions (3) are homogeneous linear equations in the coefficients of F. Under appropriate assumptions relating T, S to the degree of F, these can be solved non-trivially; and using things like Lemma 1 of Chapter 2 or Lemma 1 of Chapter 6 one can make sure that the resulting coefficients are not too large.

Next, Gelfond used analytic techniques to show that the values $f^{(t)}(s\log\alpha)$ are very small on a range larger than (3). Compare (8) of Chapter 2 and the use of Cauchy's Theorem in section 5 of Chapter 6. These values are still algebraic numbers, and then arithmetic techniques show that they are in fact zero. Compare Lemma 3 of Chapter 2 or Lemma 3 of Chapter 6 (nowadays one tends to use heights, with a definition slightly different from that of Chapter 1).

Together these lead to (3) for the new range

$$t < 2T, \quad 1 \leq s \leq 2S \tag{5}$$

slightly larger than (4) (this is not quite consistent with Baker's remark near the end of section 1 of Chapter 2 or the method of section 5 of Chapter 6, but it simplifies the proof a little).

And now the step from (4) to (5) can be iterated, and even indefinitely. This provides infinite multiplicities, and so this f must be identically zero. Taking into account the irrationality of β, we see that this implies that F is also identically zero; our required contradiction.

Next for $n = 3$ we have to reach a similar contradiction from

$$\beta_1\log\alpha_1 + \beta_2\log\alpha_2 = \log\alpha' \tag{6}$$

instead of (1), where $\alpha_1, \alpha_2, \alpha'$ are non-zero algebraic and β_1, β_2 are algebraic, this time with $1, \beta_1, \beta_2$ linearly independent over $\mathbf{Q}$. Even this was a new result.

Baker's first step looks natural: to construct now

$$f(z_1, z_2) = F(e^{z_1}, e^{z_2}, e^{\beta_1 z_1 + \beta_2 z_2}) \tag{7}$$

instead of (2) with many zeroes; but no-one had written this down before. Still less had anyone considered multiplicities, now defined by the partial derivatives

$$\left(\frac{\partial}{\partial z_1}\right)^{t_1}\left(\frac{\partial}{\partial z_2}\right)^{t_2} f(s\log\alpha_1, s\log\alpha_2) = 0 \tag{8}$$

instead of (3). Note that there is no $(s_1\log\alpha_1, s_2\log\alpha_2)$ here, because we do not have the Cartesian product situation for $\mathbf{C}^2$ mentioned in section 1 of Chapter 2. In fact our $(s\log\alpha_1, s\log\alpha_2)$ lie on a complex line in $\mathbf{C}^2$.

Baker took a range

$$t_1 + t_2 < T,\quad 1 \le s \le S \tag{9}$$

instead of (4), and the problem is then to increase this as in (5).

Now the experts know that the world of two complex variables is very different from that of a single variable. Possibly Baker did not know this. Anyway, to this day no-one knows how to reach $t_1 + t_2 < 2T$ as in (5).

He probably started by reducing to $\mathbf{C}$ via

$$g(z) = f(z\log\alpha_1, z\log\alpha_2). \tag{10}$$

Then we deduce

$$g^{(t)}(s) = 0 \quad (t < T,\ 1 \le s \le S).$$

The twin analytic-arithmetic argument then shows that $g(s) = 0$ for a larger range of s, that is, $f(s\log\alpha_1, s\log\alpha_2) = 0$. However, as it stands we cannot deduce even $g'(s) = 0$ because differentiation in (10) introduces transcendental numbers, so we cannot get at, say,

$$\left(\frac{\partial}{\partial z_1}\right) f(s\log\alpha_1, s\log\alpha_2) \tag{11}$$

in this way.

Now that we have set up the scene, Baker's solution to this problem may seem in retrospect obvious: we use (10) with f replaced by $(\partial/\partial z_1)f$. For the new g we get almost (9), but now only for $t < T - 1$. However this tiny loss does not affect the argument, and we find indeed that (11) vanishes on the larger range of s.

And what about higher derivatives? To get at some

$$\left(\frac{\partial}{\partial z_1}\right)^{\tau_1}\left(\frac{\partial}{\partial z_2}\right)^{\tau_2} f(s\log\alpha_1, s\log\alpha_2)$$

we use (10) with f replaced by $(\partial/\partial z_1)^{\tau_1}(\partial/\partial z_2)^{\tau_2} f$. We then get (9) for $t < T - \tau_1 - \tau_2$. If we aim for all τ_1, τ_2 just with $\tau_1 + \tau_2 < T$, then hardly anything remains of the multiplicity; so it is wiser to restrict to say $\tau_1 + \tau_2 < T/2$, thus securing $t < T/2$. Now the loss is less tiny, but still acceptable.

We end up with (8) on the range, say,

$$t_1 + t_2 < T/2, \quad 1 \leq s \leq 8S. \tag{12}$$

As (9) is about $T^2S/2$ conditions and (12) is about T^2S conditions, we do actually have more zeroes.

But now another problem arises: we cannot iterate indefinitely the step from (9) to (12).

In fact a related problem had turned up before Baker; for example when trying to show that the two sides of (1) cannot even be too near to each other. And indeed Baker was able to extend the classical methods; in this case the argument of section 5 of Chapter 2 amounts to the use of a non-vanishing Vandermonde determinant.

All this extends to n logarithms, and then to include 1 as in Theorem 2.1. In our notation assuming

$$\beta_0 + \beta_1 \log \alpha_1 + \cdots + \beta_{n-1} \log \alpha_{n-1} = \log \alpha', \tag{13}$$

we have to use

$$f(z_0, z_1, \ldots, z_{n-1}) = F(z_0, e^{z_1}, \ldots, e^{z_{n-1}}, e^{\beta_0 z_0 + \beta_1 z_1 + \cdots + \beta_{n-1} z_{n-1}})$$

in place of (7) at $(s, s \log \alpha_1, \ldots, s \log \alpha_{n-1})$ – compare Lemma 2 of Chapter 2.

This completes our account of Chapter 2. On the way, we have mentioned the problem of approximate versions of (1), and the corresponding generalizations to (6) and (13) are treated in Chapter 3. It is these that are needed for the applications in Chapters 4 and 5.

It is these applications, to Diophantine equations and class numbers, that were the most spectacular of his achievements. It is enough here to cite the first ever upper bounds for the solutions of Mordell's equation $y^2 = x^3 + k$ with a history going back to 1621, and the verification of Gauss's conjectures from 1801 about imaginary quadratic fields with class numbers $h = 1$ and $h = 2$.

But one should not overlook the less spectacular material in Chapter 6, whose subsequent developments (by others) will be described in the afterword.

Preface

Fermat, Euler, Lagrange, Legendre ... introitum ad penetralia huius divinae scientiae aperuerunt, quantisque divitiis abundent patefecerunt

Gauss, Disquisitiones Arithmeticae

The study of transcendental numbers, springing from such diverse sources as the ancient Greek question concerning the squaring of the circle, the rudimentary researches of Liouville and Cantor, Hermite's investigations on the exponential function and the seventh of Hilbert's famous list of 23 problems, has now developed into a fertile and extensive theory, enriching widespread branches of mathematics; and the time has seemed opportune to prepare a systematic treatise. My aim has been to provide a comprehensive account of the recent major discoveries in the field; the text includes, more especially, expositions of the latest theories relating to linear forms in the logarithms of algebraic numbers, of Schmidt's generalization of the Thue–Siegel–Roth theorem, of Shidlovsky's work on Siegel's E-functions and of Sprindžuk's solution to the Mahler conjecture. Classical aspects of the subject are discussed in the course of the narrative; in particular, to facilitate the acquisition of a true historical perspective, a survey of the theory as it existed at about the turn of the century is given at the beginning. Proofs in the subject tend, as will be appreciated, to be long and intricate, and thus it has been necessary to select for detailed treatment only the most fundamental results; moreover, generally speaking, emphasis has been placed on arguments which have led to the strongest propositions known to date or have yielded the widest application. Nevertheless, it is hoped that adequate references have been included to associated works.

Notwithstanding its long history, it will be apparent that the theory of transcendental numbers bears a youthful countenance. Many topics would certainly benefit by deeper studies and several famous longstanding problems remain open. As examples, one need mention only the celebrated conjectures concerning the algebraic independence of e and π and the transcendence of Euler's constant γ, the solution to either of which would represent a major advance. If this book should play some small rôle in promoting future progress, the author will be well satisfied.

The text has arisen from numerous lectures delivered in Cambridge, America and elsewhere, and it has also formed the substance of an Adams Prize essay.

I am grateful to Dr D. W. Masser for his kind assistance in checking the proofs, and also to the Cambridge University Press for the care they have taken with the printing.

Cambridge, 1974 A.B.

1

THE ORIGINS

1. Liouville's theorem

The theory of transcendental numbers was originated by Liouville in his famous memoir† of 1844 in which he obtained, for the first time, a class, *très-étendue*, as it was described in the title of the paper, of numbers that satisfy no algebraic equation with integer coefficients. Some isolated problems pertaining to the subject, however, had been formulated long before this date, and the closely related study of irrational numbers had constituted a major focus of attention for at least a century preceding. Indeed, by 1744, Euler had already established the irrationality of e, and, by 1761, Lambert had confirmed the irrationality of π. Moreover, the early studies of continued fractions had revealed several basic features concerning the approximation of irrational numbers by rationals. It was known, for instance, that for any irrational α there exists an infinite sequence of rationals p/q $(q > 0)$ such that‡ $|\alpha - p/q| < 1/q^2$, and it was known also that the continued fraction of a quadratic irrational is ultimately periodic, whence there exists $c = c(\alpha) > 0$ such that $|\alpha - p/q| > c/q^2$ for all rationals p/q $(q > 0)$. Liouville observed that a result of the latter kind holds more generally, and that there exists in fact a limit to the accuracy with which any algebraic number, not itself rational, can be approximated by rationals. It was this observation that provided the first practical criterion whereby transcendental numbers could be constructed.

Theorem 1.1. *For any algebraic number α with degree $n > 1$, there exists $c = c(\alpha) > 0$ such that $|\alpha - p/q| > c/q^n$ for all rationals p/q $(q > 0)$.*

The theorem follows almost at once from the definition of an algebraic number. A real or complex number is said to be algebraic if it is a zero of a polynomial with integer coefficients; every algebraic

† *C.R.* **18** (1844), 883–5, 910–11; *J. Math. pures appl.* **16** (1851), 133–42. For abbreviations see page 130.

‡ This is in fact easily verified; for any integer $Q > 1$, two of the $Q+1$ numbers 1, $\{q\alpha\}$ $(0 \leqslant q < Q)$, where $\{q\alpha\}$ denotes the fractional part of $q\alpha$, lie in one of the Q subintervals of length $1/Q$ into which $[0, 1]$ can be divided, and their difference has the form $q\alpha - p$.

number α is the zero of some such irreducible† polynomial, say P, unique up to a constant multiple, and the degree of α is defined as the degree of P. It suffices to prove the theorem when α is real; in this case, for any rational p/q $(q > 0)$, we have by the mean value theorem:

$$-P(p/q) = P(\alpha) - P(p/q) = (\alpha - p/q)\,P'(\xi)$$

for some ξ between p/q and α. Clearly one can assume that $|\alpha - p/q| < 1$, for the result would otherwise be valid trivially; then $|\xi| < 1 + |\alpha|$ and thus $|P'(\xi)| < 1/c$ for some $c = c(\alpha) > 0$; hence

$$|\alpha - p/q| > c\,|P(p/q)|.$$

But, since P is irreducible, we have $P(p/q) \neq 0$, and the integer $|q^n P(p/q)|$ is therefore at least 1; the theorem follows. Note that one can easily give an explicit value for c; in fact one can take

$$c^{-1} = n^2(1 + |\alpha|)^{n-1} H,$$

where H denotes the height of α, that is, the maximum of the absolute values of the coefficients of P.

A real or complex number that is not algebraic is said to be transcendental. In view of Theorem 1.1, an obvious instance of such a number is given by $\xi = \sum_{n=1}^{\infty} 10^{-n!}$. For if we write

$$p_j = 10^{j!} \sum_{n=1}^{j} 10^{-n!}, \qquad q_j = 10^{j!} \quad (j = 1, 2, \ldots),$$

then p_j, q_j are relatively prime rational integers and we have

$$\begin{aligned} |\xi - p_j/q_j| &= \sum_{n=j+1}^{\infty} 10^{-n!} \\ &< 10^{-(j+1)!}(1 + 10^{-1} + 10^{-2} + \ldots) = \tfrac{10}{9} q_j^{-j-1} < q_j^{-j}. \end{aligned}$$

Many other transcendental numbers can be specified on the basis of Liouville's theorem; indeed any non-terminating decimal in which there occur sufficiently long blocks of zeros, or any continued fraction in which the partial quotients increase sufficiently rapidly, provides an example. Numbers of this kind, that is real ξ which possess a sequence of distinct rational approximations p_n/q_n $(n = 1, 2, \ldots)$ such that $|\xi - p_n/q_n| < 1/q_n^{\omega_n}$, where $\limsup \omega_n = \infty$, have been termed Liouville numbers; and, of course, these are transcendental. But other,

† That is, does not factorize over the integers or, equivalently, by Gauss' lemma, over the rationals.

less obvious, applications of Liouville's idea to the construction of transcendental numbers have been described; in particular, Maillet[†] used an extension of Theorem 1.1 concerning approximations by quadratic irrationals to establish the transcendence of a remarkable class of quasi-periodic continued fractions.[‡]

In 1874, Cantor introduced the concept of countability and this leads at once to the observation that 'almost all' numbers are transcendental. Cantor's work may be regarded as the forerunner of some important metrical theory about which we shall speak in Chapter 9.

2. Transcendence of *e*

In 1873, there appeared Hermite's epoch-making memoir entitled *Sur la fonction exponentielle*[§] in which he established the transcendence of e, the natural base for logarithms. The irrationality of e had been demonstrated, as remarked earlier, by Euler in 1744, and Liouville had shown in 1840, directly from the defining series, that in fact neither e nor e^2 could be rational or a quadratic irrational; but Hermite's work began a new era. In particular, within a decade, Lindemann succeeded in generalizing Hermite's methods and, in a classic paper,[‖] he proved that π is transcendental and solved thereby the ancient Greek problem concerning the quadrature of the circle. The Greeks had sought to construct, with ruler and compasses only, a square with area equal to that of a given circle. This plainly amounts to constructing two points in the plane at a distance $\sqrt{\pi}$ apart, assuming that a unit length is prescribed. But, since all points capable of construction are defined by the intersection of lines and circles, it follows easily that their co-ordinates in a suitable frame of reference are given by algebraic numbers. Thus the transcendence of π implies that the quadrature of the circle is impossible.

The work of Hermite and Lindemann was simplified by Weierstrass[¶] in 1885, and further simplified by Hilbert,[††] Hurwitz[‡‡] and Gordan[§§] in 1893. We proceed now to demonstrate the transcendence of e and π in a style suggested by these later writers.

† See Bibliography.

‡ Cf. *Mathematika*, 9 (1962), 1–8.

§ *C.R.* 77; = *Oeuvres* III, 150–81.

‖ *M.A.* 20 (1882), 213–25.

¶ *Werke* II, 341–62.

†† *Ges. Abh.* I, 1–4.

‡‡ *Göttingen Nachrichten* (1893), 153–5.

§§ *M.A.* 43 (1893), 222–5.

Theorem 1.2. *e is transcendental.*

Preliminary to the proof, we observe that if $f(x)$ is any real polynomial with degree m, say, and if

$$I(t) = \int_0^t e^{t-u} f(u)\,du,$$

where t is an arbitrary complex number and the integral is taken over the line joining 0 and t, then, by repeated integration by parts, we have†

$$I(t) = e^t \sum_{j=0}^{m} f^{(j)}(0) - \sum_{j=0}^{m} f^{(j)}(t). \tag{1}$$

Further, if $\bar{f}(x)$ denotes the polynomial obtained from f by replacing each coefficient with its absolute value, then

$$|I(t)| \leqslant \int_0^t |e^{t-u} f(u)|\,du \leqslant |t|\, e^{|t|} \bar{f}(|t|). \tag{2}$$

Suppose now that e is algebraic, so that

$$q_0 + q_1 e + \ldots + q_n e^n = 0 \tag{3}$$

for some integers $n > 0$, $q_0 \neq 0$, $q_1, \ldots, q_n$. We shall compare estimates for

$$J = q_0 I(0) + q_1 I(1) + \ldots + q_n I(n),$$

where $I(t)$ is defined as above with

$$f(x) = x^{p-1}(x-1)^p \ldots (x-n)^p,$$

p denoting a large prime. From (1) and (3) we have

$$J = -\sum_{j=0}^{m} \sum_{k=0}^{n} q_k f^{(j)}(k),$$

where $m = (n+1)p - 1$. Now clearly $f^{(j)}(k) = 0$ if $j < p$, $k > 0$ and if $j < p-1$, $k = 0$, and thus for all j, k other than $j = p-1$, $k = 0$, $f^{(j)}(k)$ is an integer divisible by $p!$; further we have

$$f^{(p-1)}(0) = (p-1)!\,(-1)^{np}\,(n!)^p,$$

whence, if $p > n$, $f^{(p-1)}(0)$ is an integer divisible by $(p-1)!$ but not by $p!$. It follows that, if also $p > |q_0|$, then J is a non-zero integer divisible by $(p-1)!$ and thus $|J| \geqslant (p-1)!$. But the trivial estimate $\bar{f}(k) \leqslant (2n)^m$ together with (2) gives

$$|J| \leqslant |q_1|\, e\bar{f}(1) + \ldots + |q_n|\, n e^n \bar{f}(n) \leqslant c^p$$

for some c independent of p. The estimates are inconsistent if p is sufficiently large and the contradiction proves the theorem.

† $f^{(j)}(x)$ denotes the jth derivative of f.

Theorem 1.3. *π is transcendental.*

Suppose the contrary, that π is algebraic; then also $\theta = i\pi$ is algebraic. Let θ have degree d, let $\theta_1 (= \theta), \theta_2, \ldots, \theta_d$ denote the conjugates of θ and let l signify the leading coefficient in the minimal polynomial† defining θ. From Euler's equation $e^{i\pi} = -1$, we obtain

$$(1+e^{\theta_1})(1+e^{\theta_2})\ldots(1+e^{\theta_d}) = 0.$$

The product on the left can be written as a sum of 2^d terms e^{Θ}, where

$$\Theta = \epsilon_1\theta_1 + \ldots + \epsilon_d\theta_d,$$

and $\epsilon_j = 0$ or 1; we suppose that precisely n of the numbers

$$\epsilon_1\theta_1 + \ldots + \epsilon_d\theta_d$$

are non-zero, and we denote these by $\alpha_1, \ldots, \alpha_n$. We have then

$$q + e^{\alpha_1} + \ldots + e^{\alpha_n} = 0, \tag{4}$$

where q is the positive integer $2^d - n$.

We shall compare estimates for

$$J = I(\alpha_1) + \ldots + I(\alpha_n),$$

where $I(t)$ is defined as in the proof of Theorem 1.2 with

$$f(x) = l^{np}x^{p-1}(x-\alpha_1)^p \ldots (x-\alpha_n)^p,$$

p again denoting a large prime. From (1) and (4) we have

$$J = -q\sum_{j=0}^{m} f^{(j)}(0) - \sum_{j=0}^{m}\sum_{k=1}^{n} f^{(j)}(\alpha_k),$$

where $m = (n+1)p-1$. Now the sum over k is a symmetric polynomial in $l\alpha_1, \ldots, l\alpha_n$ with integer coefficients, and it follows from two applications of the fundamental theorem on symmetric functions together with the observation that each elementary symmetric function in $l\alpha_1, \ldots, l\alpha_n$ is also an elementary symmetric function in the 2^d numbers $l\Theta$, that it represents a rational integer. Further, since $f^{(j)}(\alpha_k) = 0$ when $j < p$, the latter is plainly divisible by $p!$. Clearly also $f^{(j)}(0)$ is a rational integer divisible by $p!$ when $j \neq p-1$, and

$$f^{(p-1)}(0) = (p-1)!\,(-l)^{np}\,(\alpha_1 \ldots \alpha_n)^p$$

† That is, the irreducible polynomial indicated earlier with relatively prime integer coefficients; the coefficient of x^d is called the leading coefficient, and it is assumed positive. The conjugates are the zeros of the polynomial.

is a rational integer divisible by $(p-1)!$ but not by $p!$ if p is sufficiently large. Hence, if $p > q$, we have $|J| \geqslant (p-1)!$. But from (2) we obtain

$$|J| \leqslant |\alpha_1|\, e^{|\alpha_1|}\bar{f}(|\alpha_1|)+\ldots+|\alpha_n|\, e^{|\alpha_n|}\bar{f}(|\alpha_n|) \leqslant c^p$$

for some c independent of p. The estimates are inconsistent for p sufficiently large, and the contradiction proves the theorem.

3. Lindemann's theorem

The two preceding theorems, that is the transcendence of e and π, are special cases of a much more general result which Lindemann sketched in his original memoir of 1882, and which was later rigorously demonstrated by Weierstrass.

Theorem 1.4. *For any distinct algebraic numbers $\alpha_1, \ldots, \alpha_n$ and any non-zero algebraic numbers $\beta_1, \ldots, \beta_n$ we have*

$$\beta_1 e^{\alpha_1}+\ldots+\beta_n e^{\alpha_n} \neq 0.$$

It follows at once from Theorem 1.4 that $e^{\alpha_1}, \ldots, e^{\alpha_n}$ are algebraically independent for all algebraic $\alpha_1, \ldots, \alpha_n$ linearly independent over the rationals; this form of the result is generally known as Lindemann's theorem. As further immediate corollaries of Theorem 1.4, one sees that $\cos\alpha$, $\sin\alpha$ and $\tan\alpha$ are transcendental for all algebraic $\alpha \neq 0$, and moreover $\log\alpha$ is transcendental for algebraic α not 0 or 1.

Suppose now that the theorem is false, so that

$$\beta_1 e^{\alpha_1}+\ldots+\beta_n e^{\alpha_n} = 0. \tag{5}$$

One can clearly assume, without loss of generality, that the β's are rational integers, for this can be ensured by multiplying (5) by all the expressions obtained on allowing $\beta_1, \ldots, \beta_n$ on the left to run independently through their respective conjugates and then further multiplying by a common denominator. Furthermore, one can assume that there exist integers $0 = n_0 < n_1 < \ldots < n_r = n$, such that $\alpha_{n_t+1}, \ldots, \alpha_{n_{t+1}}$ is a complete set of conjugates for each t, and

$$\beta_{n_t+1} = \ldots = \beta_{n_{t+1}}.$$

For certainly $\alpha_1, \ldots, \alpha_n$ are zeros of some polynomial with integer coefficients and degree N, say, and if $\alpha_{n+1}, \ldots, \alpha_N$ denote the remaining zeros, we have

$$\Pi(\beta_1 e^{\alpha_{k_1}}+\ldots+\beta_N e^{\alpha_{k_N}}) = 0,$$

where the product is over all permutations $k_1, \ldots, k_N$ of $1, \ldots, N$ and

$\beta_{n+1} = \ldots = \beta_N = 0$. The left-hand side can be expressed as an aggregate of terms $\exp(h_1\alpha_1 + \ldots + h_N\alpha_N)$ with integer coefficients, where $h_1, \ldots, h_N$ are integers with sum $N!$, and clearly $h_1\alpha_{k_1} + \ldots + h_N\alpha_{k_N}$ taken over all permutations $k_1, \ldots, k_N$ of $1, \ldots, N$ is a complete set of conjugates; the condition concerning the equality of the β's follows by symmetry. Note also that, after collecting terms with the same exponents, one at least of the new coefficients β will be non-zero; this is readily confirmed by considering the coefficient of the term with exponent that is highest according to the ordering of the complex numbers $z = x + iy$ given by $z_1 < z_2$ if $x_1 < x_2$ or $x_1 = x_2$ and $y_1 < y_2$.

Let now l be any positive integer such that $l\alpha_1, \ldots, l\alpha_n$ and $l\beta_1, \ldots, l\beta_n$ are algebraic integers,† and let

$$f_i(x) = l^{np}\{(x-\alpha_1)\ldots(x-\alpha_n)\}^p/(x-\alpha_i) \quad (1 \leqslant i \leqslant n),$$

where p denotes a large prime. We shall compare estimates for $|J_1 \ldots J_n|$, where

$$J_i = \beta_1 I_i(\alpha_1) + \ldots + \beta_n I_i(\alpha_n) \quad (1 \leqslant i \leqslant n),$$

and $I_i(t)$ is defined as in the proof of Theorem 1.2, with $f = f_i$. From (1) and (5) we have

$$J_i = -\sum_{j=0}^{m}\sum_{k=1}^{n} \beta_k f_i^{(j)}(\alpha_k),$$

where $m = np-1$. Further, $f_i^{(j)}(\alpha_k)$ is an algebraic integer divisible‡ by $p!$ unless $j = p-1$, $k = i$; and in the latter case we have

$$f_i^{(p-1)}(\alpha_i) = l^{np}(p-1)! \prod_{\substack{k=1 \\ k \neq i}}^{n} (\alpha_i - \alpha_k)^p,$$

so that it is an algebraic integer divisible by $(p-1)!$ but not by $p!$ if p is sufficiently large. It follows that J_i is a non-zero algebraic integer divisible by $(p-1)!$. Further, by the initial assumptions, we have

$$J_i = -\sum_{j=0}^{m}\sum_{t=0}^{r-1} \beta_{n_t+1}\{f_i^{(j)}(\alpha_{n_t+1}) + \ldots + f_i^{(j)}(\alpha_{n_{t+1}})\},$$

and here each sum over t can be expressed as a polynomial in α_i with rational coefficients independent of i; for clearly, since $\alpha_1, \ldots, \alpha_n$ is a complete set of conjugates, the coefficients of $f_i^{(j)}(x)$ can be expressed in this form. Thus $J_1 \ldots J_n$ is rational, and so in fact a rational integer

† An algebraic number is said to be an algebraic integer if the leading coefficient in its minimal defining polynomial is 1; if α is an algebraic number and l is the leading coefficient in its minimal polynomial, then $l\alpha$ is an algebraic integer.

‡ That is, the quotient is an algebraic integer.

divisible by $((p-1)!)^n$. Hence we have $|J_1 \dots J_n| \geqslant (p-1)!$. But (2) gives

$$|J_i| \leqslant \sum_{k=1}^{n} |\alpha_k \beta_k| \, e^{|\alpha_k|} \bar{f}(|\alpha_k|) \leqslant c^p,$$

for some c independent of p, and the inequalities are inconsistent if p is sufficiently large. The contradiction proves the theorem.

The above proofs are simplified versions of the original arguments of Hermite and Lindemann and their motivation may seem obscure; indeed there is no explanation *a priori* for the introduction of the functions I and f. A deeper insight can best be obtained by studying the basic memoir of Hermite where, in modified form, the functions first occurred, but it may be said that they relate to generalizations, concerning simultaneous approximation, of the convergents in the continued fraction expansion of e^x. Further light on the topic will be shed by Chapters 10 and 11. Lindemann's theorem formed the summit of the accomplishments of the last century, and our survey of the period is herewith concluded.

2

LINEAR FORMS IN LOGARITHMS

1. Introduction

In 1900, at the International Congress of Mathematicians held in Paris, Hilbert raised, as the seventh of his famous list of 23 problems, the question whether an irrational logarithm of an algebraic number to an algebraic base is transcendental. The question is capable of various alternative formulations; thus one can ask whether an irrational quotient of natural logarithms of algebraic numbers is transcendental, or whether α^β is transcendental for any algebraic number $\alpha \neq 0, 1$ and any algebraic irrational β. A special case relating to logarithms of rational numbers can be traced to the writings of Euler more than a century before, but no apparent progress had been made towards its solution. Indeed, Hilbert expressed the opinion that the resolution of the problem lay farther in the future than a proof of the Riemann hypothesis or Fermat's last theorem.

The first significant advance was made by Gelfond† in 1929. Employing interpolation techniques of the kind that he had utilized previously in researches on integral integer-valued functions,‡ Gelfond showed that the logarithm of an algebraic number to an algebraic base cannot be an imaginary quadratic irrational, that is, α^β is transcendental for any algebraic number $\alpha \neq 0, 1$ and any imaginary quadratic irrational β; in particular, this implies that $e^\pi = (-1)^{-i}$ is transcendental. The result was extended to real quadratic irrationals β by Kuzmin§ in 1930. But it was clear that direct appeal to an interpolation series for $e^{\beta z}$, on which the Gelfond–Kuzmin work was based, was not appropriate for more general β, and further progress awaited a new idea. The search for the latter was concluded successfully by Gelfond‖ and Schneider¶ independently in 1934. The arguments they discovered were applicable for any irrational β and, though differing in detail, both depended on the construction of an auxiliary function that vanished at certain selected points. A similar technique had been used a few years earlier by Siegel in the course of investigations on the

† *C.R.* **189** (1929), 1224–8.

‡ *Tôhoku Math. J.* **30** (1929), 280–5.

§ *I.A.N.* **3** (1930), 583–97.

‖ *D.A.N.* **2** (1934), 1–6; *I.A.N.* **7** (1934), 623–4.

¶ *J.M.* **172** (1934), 65–9.

Bessel functions.† Herewith Hilbert's seventh problem was finally solved.

The Gelfond–Schneider theorem shows that for any non-zero algebraic numbers α_1, α_2, β_1, β_2, with $\log\alpha_1$, $\log\alpha_2$ linearly independent over the rationals, we have

$$\beta_1\log\alpha_1+\beta_2\log\alpha_2 \neq 0.$$

It was natural to conjecture that an analogous theorem would hold for arbitrarily many logarithms of algebraic numbers, and, moreover, it was soon realized that such a result would be capable of wide application. The conjecture was proved by the author‡ in 1966, and the demonstration will be the subject of the present chapter.

Theorem 2.1. *If $\alpha_1, \ldots, \alpha_n$ are non-zero algebraic numbers such that§ $\log\alpha_1, \ldots, \log\alpha_n$ are linearly independent over the rationals, then 1, $\log\alpha_1, \ldots, \log\alpha_n$ are linearly independent over the field of all algebraic numbers.*

The proof depends on the construction of an auxiliary function of several complex variables which generalizes the function of a single variable employed originally by Gelfond. Functions of several variables were utilized by Schneider‖ in his studies concerning Abelian integrals but, for many years, there appeared to be severe limitations to their serviceability in wider settings. The main difficulty concerned the basic interpolation techniques. Work in this connexion had hitherto always involved an extension in the order of the derivatives while leaving the points of interpolation fixed; however, when dealing with functions of several variables, this type of argument requires that the points in question form a cartesian product, a condition that can apparently be satisfied only with respect to particular multiply-periodic functions. The proof of Theorem 2.1 involves an extrapolation procedure, special to the present context, in which the range of interpolation is now extended while the order of the derivatives is reduced. Refinements and generalizations will be discussed in the next chapter and applications of the results to various branches of number theory will be the theme of Chapters 4 and 5.

† *Abh. Preuss Akad. Wiss.* (1929), No. 1; cf. ch. 11.
‡ *Mathematika*, **13** (1966), 204–16; **14** (1967), 102–7, 220–8.
§ Here the logarithms can take any fixed values.
‖ *J.M.* **183** (1941), 110–28.

2. Corollaries

Before proceeding to the proof of Theorem 2.1, we record a few immediate corollaries.

Theorem 2.2. *Any non-vanishing linear combination of logarithms of algebraic numbers with algebraic coefficients is transcendental.*

In other words, for any non-zero algebraic numbers $\alpha_1, \ldots, \alpha_n$ and any algebraic numbers $\beta_0, \beta_1, \ldots, \beta_n$ with $\beta_0 \neq 0$ we have

$$\beta_0 + \beta_1 \log \alpha_1 + \ldots + \beta_n \log \alpha_n \neq 0.$$

This plainly holds for $n = 0$. We assume the validity for $n < m$, where m is a positive integer, and proceed to prove the proposition for $n = m$. Now if $\log \alpha_1, \ldots, \log \alpha_m$ are linearly independent over the rationals then the result follows from Theorem 2.1. Thus we can suppose that there exist rationals $\rho_1, \ldots, \rho_m$, with say $\rho_r \neq 0$, such that

$$\rho_1 \log \alpha_1 + \ldots + \rho_m \log \alpha_m = 0.$$

Clearly we have

$$\rho_r(\beta_0 + \beta_1 \log \alpha_1 + \ldots + \beta_m \log \alpha_m) = \beta_0' + \beta_1' \log \alpha_1 + \ldots + \beta_m' \log \alpha_m,$$

where $$\beta_0' = \rho_r \beta_0, \quad \beta_j' = \rho_r \beta_j - \rho_j \beta_r \quad (1 \leqslant j \leqslant m),$$

and also $\beta_0' \neq 0$, $\beta_r' = 0$; the required result follows by induction.

Theorem 2.3. *$e^{\beta_0}\alpha_1^{\beta_1} \ldots \alpha_n^{\beta_n}$ is transcendental for any non-zero algebraic numbers $\alpha_1, \ldots, \alpha_n, \beta_0, \beta_1, \ldots, \beta_n$.*

Indeed, if $\alpha_{n+1} = e^{\beta_0}\alpha_1^{\beta_1} \ldots \alpha_n^{\beta_n}$ were algebraic, then

$$\beta_1 \log \alpha_1 + \ldots + \beta_n \log \alpha_n - \log \alpha_{n+1} \quad (= -\beta_0)$$

would be algebraic and non-vanishing, contrary to Theorem 2.2. There is a natural analogue to Theorem 2.3 in the case $\beta_0 = 0$:

Theorem 2.4. *$\alpha_1^{\beta_1} \ldots \alpha_n^{\beta_n}$ is transcendental for any algebraic numbers $\alpha_1, \ldots, \alpha_n$, other than 0 or 1, and any algebraic numbers $\beta_1, \ldots, \beta_n$ with $1, \beta_1, \ldots, \beta_n$ linearly independent over the rationals.*

For the proof, it suffices to show that for any algebraic numbers $\alpha_1, \ldots, \alpha_n$, other than 0 or 1, and any algebraic numbers $\beta_1, \ldots, \beta_n$, linearly independent over the rationals, we have

$$\beta_1 \log \alpha_1 + \ldots + \beta_n \log \alpha_n \neq 0;$$

in fact the theorem follows on applying this with n replaced by $n+1$ and $\beta_{n+1} = -1$. The proposition plainly holds for $n = 1$; we assume the validity for $n < m$, where m is a positive integer, and proceed to prove the assertion for $n = m$. The result is an immediate consequence of Theorem 2.1 if $\log \alpha_1, \ldots, \log \alpha_n$ are linearly independent over the rationals; thus we can suppose that there exist rationals $\rho_1, \ldots, \rho_m$ and numbers β'_j as in the proof of Theorem 2.2, with now $\beta_0 = \beta'_0 = 0$. It is clear that if $\beta_1, \ldots, \beta_m$ are linearly independent over the rationals, then so also are the β'_j, with j not 0 or r, and the theorem follows by induction.

Finally, from particular cases of the above theorems, it is evident that $\pi + \log \alpha$ is transcendental for any algebraic number $\alpha \neq 0$ (which includes the transcendence of π) and that $e^{\alpha\pi+\beta}$ is transcendental for any algebraic numbers α, β with $\beta \neq 0$ (which includes the transcendence of e).

3. Notation

The remainder of the chapter is devoted to a proof of Theorem 2.1. We suppose that the theorem is false, so that there exist algebraic numbers $\beta_0, \beta_1, \ldots, \beta_n$, not all 0, such that

$$\beta_0 + \beta_1 \log \alpha_1 + \ldots + \beta_n \log \alpha_n = 0,$$

and we ultimately derive a contradiction. Clearly one at least of $\beta_1, \ldots, \beta_n$ is not 0 and, without loss of generality, we can suppose that $\beta_n \neq 0$. Since the above equation continues to hold with $\beta'_j = -\beta_j/\beta_n$ in place of β_j, we can further suppose, without loss of generality, that $\beta_n = -1$; we have then

$$e^{\beta_0} \alpha_1^{\beta_1} \ldots \alpha_{n-1}^{\beta_{n-1}} = \alpha_n. \tag{1}$$

We denote by $c, c_1, c_2, \ldots$ positive numbers which depend only on the α's, β's and the original determinations of the logarithms. By h we signify a positive integer which exceeds a sufficiently large number c as above.

We note, for later reference, that if α is any algebraic number satisfying

$$A_0 \alpha^d + A_1 \alpha^{d-1} + \ldots + A_d = 0,$$

where $A_0, \ldots, A_d$ are rational integers with absolute values at most A, then, for each non-negative integer j, we have

$$(A_0 \alpha)^j = A_0^{(j)} + A_1^{(j)} \alpha + \ldots + A_{d-1}^{(j)} \alpha^{d-1}$$

for some rational integers $A_m^{(j)}$ with absolute values at most $(2A)^j$; this is an obvious consequence of the recurrence relations

$$A_m^{(j)} = A_0 A_{m-1}^{(j-1)} - A_{d-m} A_{d-1}^{(j-1)} \quad (0 \leqslant m < d, j \geqslant d),$$

where $A_{-1}^{(j-1)} = 0$. It follows that if d is the maximum of the degrees of $\alpha_1, \ldots, \alpha_n, \beta_0, \ldots, \beta_{n-1}$ and if $a_1, \ldots, a_n, b_0, \ldots, b_{n-1}$ are the leading coefficients in their respective minimal polynomials, then

$$(a_r \alpha_r)^j = \sum_{s=0}^{d-1} a_{rs}^{(j)} \alpha_r^s, \quad (b_r \beta_r)^j = \sum_{t=0}^{d-1} b_{rt}^{(j)} \beta_r^t, \tag{2}$$

where the $a_{rs}^{(j)}, b_{rt}^{(j)}$ are rational integers with absolute values at most c_1^j.

For brevity we shall put

$$f_{m_0, \ldots, m_{n-1}}(z_0, \ldots, z_{n-1}) = (\partial/\partial z_0)^{m_0} \ldots (\partial/\partial z_{n-1})^{m_{n-1}} f(z_0, \ldots, z_{n-1}),$$

where f denotes an integral function and $m_0, \ldots, m_{n-1}$ are non-negative integers.

4. The auxiliary function

Our purpose now is to describe the auxiliary function Φ that is fundamental to the proof of Theorem 2.1; it is constructed in Lemma 2 below after a preliminary result on linear equations obtained by Dirichlet's box principle. Basic estimates relating to Φ are established in Lemma 3 and these are then employed for the extrapolation algorithm. Two further supplementary results are given by Lemmas 6 and 7; the former exhibits a simple, but useful, lower bound for a linear form in logarithms, and the latter furnishes a special augmentative polynomial. It will be seen that the inclusion of the 1 in the enunciation of Theorem 2.1, which yields the algebraic powers of e in the corollaries, entails a relatively large amount of additional complexity in the proof; in particular the final lemma is required essentially to deal with this feature.

Lemma 1. *Let M, N denote integers with $N > M > 0$ and let*

$$u_{ij} \quad (1 \leqslant i \leqslant M,\ 1 \leqslant j \leqslant N)$$

denote integers with absolute values at most U $(\geqslant 1)$. Then there exist integers $x_1, \ldots, x_N$ not all 0, with absolute values at most $(NU)^{M/(N-M)}$ such that

$$\sum_{j=1}^{N} u_{ij} x_j = 0 \quad (1 \leqslant i \leqslant M). \tag{3}$$

Proof. We put $B = [(NU)^{M/(N-M)}]$, where, as later, $[x]$ denotes the integral part of x. There are $(B+1)^N$ different sets of integers $x_1, \ldots, x_N$ with $0 \leqslant x_j \leqslant B$ $(1 \leqslant j \leqslant N)$, and for each such set we have

$$-V_i B \leqslant y_i \leqslant W_i B \quad (1 \leqslant i \leqslant M),$$

where y_i denotes the left-hand side of (3), and $-V_i$, W_i denote the sum of the negative and positive u_{ij} $(1 \leqslant j \leqslant N)$ respectively. Since $V_i + W_i \leqslant NU$, there are at most $(NUB+1)^M$ different sets $y_1, \ldots, y_M$. Now $(B+1)^{N-M} > (NU)^M$ and so $(B+1)^N > (NUB+1)^M$. Hence there are two distinct sets $x_1, \ldots, x_N$ which correspond to the same set $y_1, \ldots, y_M$, and their difference gives the required solution of (3).

Lemma 2. *There are integers $p(\lambda_0, \ldots, \lambda_n)$, not all 0, with absolute values at most e^{h^3}, such that the function*

$$\Phi(z_0, \ldots, z_{n-1}) = \sum_{\lambda_0=0}^{L} \cdots \sum_{\lambda_n=0}^{L} p(\lambda_0, \ldots, \lambda_n)\, z_0^{\lambda_0} e^{\lambda_n \beta_0 z_0} \alpha_1^{\gamma_1 z_1} \ldots \alpha_{n-1}^{\gamma_{n-1} z_{n-1}},$$

where $\gamma_r = \lambda_r + \lambda_n \beta_r$ $(1 \leqslant r < n)$ and $L = [h^{2-1/(4n)}]$, satisfies

$$\Phi_{m_0, \ldots, m_{n-1}}(l, \ldots, l) = 0 \tag{4}$$

for all integers l with $1 \leqslant l \leqslant h$ and all non-negative integers $m_0, \ldots, m_{n-1}$ with $m_0 + \ldots + m_{n-1} \leqslant h^2$.

Proof. It suffices, in view of (1), to determine the $p(\lambda_0, \ldots, \lambda_n)$ such that

$$\sum_{\lambda_0=0}^{L} \cdots \sum_{\lambda_n=0}^{L} p(\lambda_0, \ldots, \lambda_n)\, q(\lambda_0, \lambda_n, l)\, \alpha_1^{\lambda_1 l} \ldots \alpha_n^{\lambda_n l} \gamma_1^{m_1} \ldots \gamma_{n-1}^{m_{n-1}} = 0 \tag{5}$$

for the above ranges of $l, m_0, \ldots, m_{n-1}$, where

$$q(\lambda_0, \lambda_n, z) = \sum_{\mu_0=0}^{m_0} \binom{m_0}{\mu_0} \lambda_0(\lambda_0 - 1) \ldots (\lambda_0 - \mu_0 + 1)\, (\lambda_n \beta_0)^{m_0 - \mu_0} z^{\lambda_0 - \mu_0}.$$

On multiplying (5) by

$$P' = (a_1 \ldots a_n)^{Ll} b_0^{m_0} \ldots b_{n-1}^{m_{n-1}}, \tag{6}$$

writing

$$\gamma_r^{m_r} = \sum_{\mu_r=0}^{m_r} \binom{m_r}{\mu_r} \lambda_r^{m_r - \mu_r} (\lambda_n \beta_r)^{\mu_r},$$

and substituting from (2) for the powers of $a_r \alpha_r$ and $b_r \beta_r$ which result, we obtain

$$\sum_{s_1=0}^{d-1} \cdots \sum_{s_n=0}^{d-1} \sum_{t_0=0}^{d-1} \cdots \sum_{t_{n-1}=0}^{d-1} A(s,t)\, \alpha_1^{s_1} \ldots \alpha_n^{s_n} \beta_0^{t_0} \ldots \beta_{n-1}^{t_{n-1}} = 0,$$

where

$$A(s,t) = \sum_{\lambda_0=0}^{L} \cdots \sum_{\lambda_n=0}^{L} \sum_{\mu_0=0}^{m_0} \cdots \sum_{\mu_{n-1}=0}^{m_{n-1}} p(\lambda_0, \ldots, \lambda_n)\, q' q'' q''',$$

and q', q'', q''' are given by

$$q' = \prod_{r=1}^{n} \{a_r^{(L-\lambda_r)l} a_{r,s_r}^{(\lambda_r l)}\},$$

$$q'' = \prod_{r=1}^{n-1} \left\{\binom{m_r}{\mu_r} (b_r \lambda_r)^{m_r-\mu_r} \lambda_n^{\mu_r} b_{r,t_r}^{(\mu_r)}\right\},$$

$$q''' = \binom{m_0}{\mu_0} \lambda_0(\lambda_0-1)\dots(\lambda_0-\mu_0+1)\,\lambda_n^{m_0-\mu_0} b_n^{\mu_0} l^{\lambda_0-\mu_0} b_{0,t_0}^{(m_0-\mu_0)}.$$

Thus (4) will be satisfied if the d^{2n} equations $A(s,t) = 0$ hold. Now these represent linear equations in the $p(\lambda_0, \dots, \lambda_n)$ with integer coefficients. Since $l \leqslant h$ and $\binom{m_r}{\mu_r} \leqslant 2^{m_r}$, we have

$$|q'| \leqslant \prod_{r=1}^{n} \{a_r^{(L-\lambda_r)l} c_1^{\lambda_r l}\} \leqslant c_2^{Lh},$$

$$|q''| \leqslant \prod_{r=1}^{n-1} (c_3 L)^{m_r},$$

$$|q'''| \leqslant 2^{m_0}(\lambda_0 b_n)^{\mu_0} (c_1 \lambda_n)^{m_0-\mu_0} l^{\lambda_0-\mu_0} \leqslant (c_3 L)^{m_0} h^L,$$

and, by virtue of the inequalities

$$(m_0+1)\dots(m_{n-1}+1) \leqslant 2^{m_0+\dots+m_{n-1}} \leqslant 2^{h^2},$$

it follows easily that the coefficient of $p(\lambda_0, \dots, \lambda_n)$ in the linear form $A(s,t)$, namely

$$\sum_{\mu_0=0}^{m_0} \dots \sum_{\mu_{n-1}=0}^{m_{n-1}} q'q''q''',$$

has absolute value at most $U = (2c_3 L)^{h^2} c_4^{Lh}$. Further, there are at most $h(h^2+1)^n$ distinct sets of integers $l, m_0, \dots, m_{n-1}$, and hence there are $M \leqslant d^{2n}h(h^2+1)^n$ equations $A(s,t) = 0$ corresponding to them. Furthermore, there are $N = (L+1)^{n+1}$ unknowns $p(\lambda_0, \dots, \lambda_n)$ and we have

$$N > h^{(2-1/(4n))(n+1)} \geqslant h^{2n+\frac{5}{4}} > 2d^{2n}h(h^2+1)^n \geqslant 2M.$$

Thus, by Lemma 1, the equations can be solved non-trivially and the integers $p(\lambda_0, \dots, \lambda_n)$ can be chosen to have absolute values at most

$$NU \leqslant h^{2n+2}(2c_3 L)^{h^2} c_4^{Lh} \leqslant e^{h^3}$$

if h is sufficiently large, as required.

Lemma 3. *Let $m_0, \dots, m_{n-1}$ be any non-negative integers with*

$$m_0 + \dots + m_{n-1} \leqslant h^2,$$

and let

$$f(z) = \Phi_{m_0,\dots,m_{n-1}}(z, \dots, z). \tag{7}$$

Then, for any number z, we have $|f(z)| \leqslant c_6^{h^3+L|z|}$. Further, for any positive integer l, either $f(l) = 0$ or $|f(l)| > c_6^{-h^3-Ll}$.

Proof. The function $f(z)$ is given by

$$P \sum_{\lambda_0=0}^{L} \cdots \sum_{\lambda_n=0}^{L} p(\lambda_0, \ldots, \lambda_n)\, q(\lambda_0, \lambda_n, z)\, \alpha_1^{\lambda_1 z} \ldots \alpha_n^{\lambda_n z} \gamma_1^{m_1} \ldots \gamma_{n-1}^{m_{n-1}},$$

where $q(\lambda_0, \lambda_n, z)$ is defined in Lemma 2 and

$$P = (\log \alpha_1)^{m_1} \ldots (\log \alpha_{n-1})^{m_{n-1}}.$$

We have

$$|q(\lambda_0, \lambda_n, z)| \leqslant (c_7 L)^{m_0} (|z| + 1)^L \sum_{\mu_0=0}^{m_0} \binom{m_0}{\mu_0} = (2c_7 L)^{m_0} (|z| + 1)^L,$$

$$|\alpha_1^{\lambda_1 z} \ldots \alpha_n^{\lambda_n z}| \leqslant c_8^{L|z|}, \quad |P\gamma_1^{m_1} \ldots \gamma_{n-1}^{m_{n-1}}| \leqslant (c_9 L)^{m_1+\ldots+m_{n-1}},$$

and the number of terms in the above multiple sum is at most h^{2n+2}; the required estimate for $|f(z)|$ now follows by virtue of the inequalities

$$L \leqslant h^2, \quad m_0 + \ldots + m_{n-1} \leqslant h^2, \quad |p(\lambda_0, \ldots, \lambda_n)| \leqslant e^{h^3}.$$

To prove the second assertion, we begin by noting that the number $f' = (P'/P) f(l)$, where P' is defined by (6), is an algebraic integer with degree at most d^{2n}. Further, by estimates as above, we see that any conjugate of f', obtained by substituting arbitrary conjugates for the α_r, β_r, has absolute value at most $c_{10}^{h^3+Ll}$; and clearly the same bound obtains for P'/P. But if $f' \neq 0$, then the norm† of f' has absolute value at least 1 and so

$$|f'| \geqslant c_{10}^{-(h^3+Ll)d^{2n}}.$$

This gives the required result.

Lemma 4. *Let J be any integer satisfying $0 \leqslant J \leqslant (8n)^2$. Then (4) holds for all integers l with $1 \leqslant l \leqslant h^{1+J/(8n)}$ and all non-negative integers $m_0, \ldots, m_{n-1}$ with $m_0 + \ldots + m_{n-1} \leqslant h^2/2^J$.*

Proof. The result holds for $J = 0$ by Lemma 2. Let K be an integer with $0 \leqslant K < (8n)^2$ and assume that the lemma is valid for

$$J = 0, 1, \ldots, K.$$

We proceed to prove the proposition for $J = K + 1$.

† The product of the conjugates; it is plainly a rational integer.

It suffices to show that for any integer l with $R_K < l \leqslant R_{K+1}$ and any set of non-negative integers $m_0, \ldots, m_{n-1}$ with

$$m_0 + \ldots + m_{n-1} \leqslant S_{K+1},$$

we have $f(l) = 0$, where $f(z)$ is defined by (7) and

$$R_J = [h^{1+J/(8n)}], \quad S_J = [h^2/2^J] \quad (J = 0, 1, \ldots).$$

By the inductive hypothesis we see that $f_m(r) = 0$ for all integers r, m with $1 \leqslant r \leqslant R_K$, $0 \leqslant m \leqslant S_{K+1}$; for clearly $f_m(r)$ is given by

$$(\partial/\partial z_0 + \ldots + \partial/\partial z_{n-1})^m \, \Phi_{m_0, \ldots, m_{n-1}}(z_0, \ldots, z_{n-1}),$$

evaluated at the point $z_0 = \ldots = z_{n-1} = r$, that is by

$$\Sigma m!(j_0! \ldots j_{n-1}!)^{-1} \, \Phi_{m_0+j_0, \ldots, m_{n-1}+j_{n-1}}(r, \ldots, r),$$

where the sum is over all non-negative integers $j_0, \ldots, j_{n-1}$ with $j_0 + \ldots + j_{n-1} = m$, and the derivatives here are 0 since

$$m_0 + \ldots + m_{n-1} + j_0 + \ldots + j_{n-1} \leqslant 2S_{K+1} \leqslant S_K.$$

Thus $f(z)/F(z)$, where

$$F(z) = \{(z-1) \ldots (z - R_K)\}^{S_{K+1}},$$

is regular within and on the circle C with centre the origin and radius $R = R_{K+1} h^{1/(8n)}$, and hence, by the maximum-modulus principle,

$$\theta |F(l)| \geqslant \Theta |f(l)|, \tag{8}$$

where θ, Θ denote respectively the upper bound of $|f(z)|$ and the lower bound of $|F(z)|$ with z on C. Now clearly $\Theta \geqslant (\frac{1}{2}R)^{R_K S_{K+1}}$ and, by Lemma 3, $\theta \leqslant c_5^{h^3+LR}$. Further, we have $|F(l)| \leqslant R_{K+1}^{R_K S_{K+1}}$ and, by Lemma 3 again, either $f(l) = 0$ or $|f(l)| > c_6^{-h^3-LR}$. But, in view of (8), the latter possibility gives

$$(c_5 c_6)^{h^3+LR} \geqslant (\tfrac{1}{2}h^{1/(8n)})^{R_K S_{K+1}},$$

and, since $K < (8n)^2$ and

$$LR \leqslant h^{3+K/(8n)} \leqslant 2^{K+3} R_K S_{K+1},$$

the inequality is untenable if h is sufficiently large. Hence $f(l) = 0$, and the lemma follows by induction.

Lemma 5. *Writing $\phi(z) = \Phi(z, \ldots, z)$, we have*

$$|\phi_j(0)| < \exp(-h^{8n}) \quad (0 \leqslant j \leqslant h^{8n}). \tag{9}$$

Proof. From Lemma 4 we see that (4) holds for all integers l and non-negative integers $m_0, \ldots, m_{n-1}$ satisfying $1 \leqslant l \leqslant X$ and

$$m_0 + \ldots + m_{n-1} \leqslant Y,$$

where $X = h^{8n}$ and $Y = [h^2/2^{(8n)^2}]$. Hence, as in the proof of Lemma 4, we obtain $\phi_m(r) = 0$ for all integers r, m with $1 \leqslant r \leqslant X$, $0 \leqslant m \leqslant Y$. It follows that $\phi(z)/E(z)$, where

$$E(z) = \{(z-1) \ldots (z-X)\}^Y,$$

is regular within and on the circle Γ with centre the origin and radius $R = Xh^{1/(8n)}$, and so, by the maximum-modulus principle, we have, for any w with $|w| < X$,

$$|\phi(w)| \leqslant \xi \Xi^{-1} |E(w)|,$$

where ξ and Ξ denote respectively the upper bound of $|\phi(z)|$ and the lower bound of $|E(z)|$ with z on Γ. Now clearly

$$|E(w)| \leqslant (2X)^{XY}, \quad |\Xi| \geqslant (\tfrac{1}{2}R)^{XY},$$

and, by Lemma 3, $\xi \leqslant c_5^{h^3+LR}$. Hence we obtain

$$|\phi(w)| \leqslant c_5^{h^3+LR}(\tfrac{1}{4}h^{1/(8n)})^{-XY},$$

and since

$$LR \leqslant h^{8n+2} \leqslant 2^{(8n)^2+1}XY,$$

it follows that $|\phi(w)| < e^{-XY}$. Further, by Cauchy's formulae, we have

$$\phi_j(0) = \frac{j!}{2\pi i}\int_\Lambda \frac{\phi(w)}{w^{j+1}}\,dw,$$

where Λ denotes the circle $|w| = 1$ described in the positive sense, and the expression on the right has absolute value at most $j^j e^{-XY}$. The required estimate (9) follows at once.

Lemma 6. *For any integers $t_1, \ldots, t_n$, not all 0, and with absolute values at most T, we have*

$$|t_1 \log \alpha_1 + \ldots + t_n \log \alpha_n| > c_{11}^{-T}.$$

Proof. Let a_j $(1 \leqslant j \leqslant n)$ be the leading coefficient in the minimal defining polynomial of α_j or α_j^{-1} according as $t_j \geqslant 0$ or $t_j < 0$. Then

$$\omega = a_1^{|t_1|} \ldots a_n^{|t_n|}(\alpha_1^{t_1} \ldots \alpha_n^{t_n} - 1)$$

is an algebraic integer with degree at most d^n, and any conjugate of ω, obtained by substituting arbitrary conjugates for $\alpha_1, \ldots, \alpha_n$, has absolute value at most c_{12}^T. If $\omega = 0$ then

$$\Omega = t_1 \log \alpha_1 + \ldots + t_n \log \alpha_n$$

is a multiple of $2\pi i$, and in fact a non-zero multiple since $\log \alpha_1, \ldots, \log \alpha_n$ are, by hypothesis, linearly independent over the rationals; hence, in this case, the lemma is valid trivially. Otherwise the norm of ω has absolute value at least 1 and thus $|\omega| \geqslant c_{12}^{-Td^n}$. But since, for any z, $|e^z - 1| \leqslant |z| e^{|z|}$, we obtain $|\omega| \leqslant |\Omega| e^{|\Omega|} c_{13}^T$ and hence, assuming, as we may, that $|\Omega| < 1$, the lemma follows.

Lemma 7. *Let R, S be positive integers and let $\sigma_0, \ldots, \sigma_{R-1}$ be distinct complex numbers. Define σ as the maximum of 1, $|\sigma_0|, \ldots, |\sigma_{R-1}|$ and define ρ as the minimum of 1 and the $|\sigma_i - \sigma_j|$ with $0 \leqslant i < j < R$. Then, for any integers r, s with $0 \leqslant r < R$, $0 \leqslant s < S$, there exist complex numbers w_i $(0 \leqslant i < RS)$ with absolute values at most $(8\sigma/\rho)^{RS}$ such that the polynomial*

$$W(z) = \sum_{j=0}^{RS-1} w_j z^j$$

satisfies $W_j(\sigma_i) = 0$ for all i, j with $0 \leqslant i < R$, $0 \leqslant j < S$ other than $i = r, j = s$, and $W_s(\sigma_r) = 1$.

Proof. The required polynomial is given by

$$W(z) = \left(\frac{-1}{s!}\right) \frac{1}{2\pi i} \int_{C_r} \frac{(\zeta - \sigma_r)^s \, U(z)}{(\zeta - z) \, U(\zeta)} \, d\zeta,$$

where
$$U(z) = \{(z - \sigma_0) \ldots (z - \sigma_{R-1})\}^S$$

and C_r denotes a circle described in the positive sense with centre σ_r and sufficiently small radius, less than, say, ρ and $|z - \sigma_r|$ for $z \neq \sigma_r$. The proof depends on two alternative expressions for $W(z)$. First, since the absolute value of the integrand multiplied by $|\zeta|$ decreases to 0 as $|\zeta| \to \infty$ we have, by Cauchy's residue theorem,

$$W(z) = \frac{(z - \sigma_r)^s}{s!} + \frac{U(z)}{s!} \frac{1}{2\pi i} \sum_{\substack{j=0 \\ j \neq r}}^{R-1} \int_{C_j} \frac{(\zeta - \sigma_r)^s}{(\zeta - z) \, U(\zeta)} \, d\zeta,$$

where C_j, like C_r above, is a circle about σ_j with sufficiently small radius. Clearly the sum over j is a rational function of z, regular at $z = \sigma_r$ and, since $U(z)$ has a zero at $z = \sigma_r$ of order S, it follows that $W_j(\sigma_r) = 1$ if $j = s$ and 0 otherwise.

On the other hand, from Cauchy's formulae we obtain

$$W(z) = \frac{-1}{s!t!}\left[\frac{d^t}{d\zeta^t}\frac{(\zeta-\sigma_r)^S U(z)}{(\zeta-z)\,U(\zeta)}\right]_{\zeta=\sigma_r},$$

where $t = S-s-1$, and thus

$$W(z) = (-1)^{t-1}(s!)^{-1}\,U(z)\,\Sigma v(j_0,\ldots,j_{R-1})\,(\sigma_r-z)^{-j_r-1},$$

where the sum is over all non-negative integers $j_0,\ldots,j_{R-1}$ with $j_0+\ldots+j_{R-1} = t$, and

$$v(j_0,\ldots,j_{R-1}) = \prod_{\substack{i=0\\ i\neq r}}^{R-1}\binom{S+j_i-1}{j_i}(\sigma_r-\sigma_i)^{-S-j_i}.$$

Now j_r+1 lies between 1 and S inclusive and so obviously $W(z)$ is a polynomial with degree at most $RS-1$. Further, we see that $W(z)$, like $U(z)$, has a zero at $z = \sigma_i$ $(i \neq r)$ of order S, and so $W_j(\sigma_i) = 0$ for all $j < S$. Furthermore, it is clear that the typical factor in the product defining v has absolute value at most $2^{S+j_i-1}\rho^{-S-j_i}$, and thus

$$|v(j_0,\ldots,j_{R-1})| \leqslant (2/\rho)^{(R-1)S+j_0+\ldots+j_{R-1}} \leqslant (2/\rho)^{RS}.$$

On noting that the coefficients of $(\sigma_r-z)^{-j_r-1}\,U(z)$ have absolute values at most $(\sigma+1)^{RS}$ and observing, in addition, that the number of terms in the above sum does not exceed S^R, it follows easily that the coefficients of $W(z)$ have absolute values at most

$$S^R(\sigma+1)^{RS}(2/\rho)^{RS} \leqslant (8\sigma/\rho)^{RS},$$

and this completes the proof of the lemma.

5. Proof of main theorem

We proceed to show that the inequalities (9) obtained in Lemma 5 cannot all be valid, and the contradiction will establish Theorem 2.1.

We begin by writing $S = L+1$, $R = S^n$, and noting that any integer i with $0 \leqslant i < RS$ can be expressed uniquely in the form

$$i = \lambda_0+\lambda_1 S+\ldots+\lambda_n S^n,$$

where $\lambda_0,\ldots,\lambda_n$ denote integers between 0 and L inclusive. For each such i we define

$$\nu_i = \lambda_0,\quad p_i = p(\lambda_0,\ldots,\lambda_n),$$

and we put

$$\psi_i = \lambda_1\log\alpha_1+\ldots+\lambda_n\log\alpha_n.$$

Then clearly

$$\phi(z) = \sum_{i=0}^{RS-1} p_i z^{\nu_i} e^{\psi_i z}. \qquad (10)$$

Further, from Lemma 6, any two ψ_i which correspond to distinct sets $\lambda_1, \ldots, \lambda_n$ differ by at least c_{11}^{-L}; in particular, exactly R of the ψ_i are distinct, and we denote the different values, in some order, by $\sigma_0, \ldots, \sigma_{R-1}$. If σ, ρ are defined as in Lemma 7, we have then $\sigma \leqslant c_{14}L$ and $\rho \geqslant c_{15}^{-L}$.

Let now t be any suffix such that $p_t \neq 0$, let $s = \nu_t$, let r be that suffix for which $\psi_t = \sigma_r$, and let $W(z)$ denote the polynomial given by Lemma 7. By the properties of $W(z)$ specified in the lemma, we see that

$$p_t = \sum_{i=0}^{RS-1} p_i W_{\nu_i}(\psi_i).$$

Further, by Leibnitz's theorem, we have

$$W_{\nu_i}(\psi_i) = \sum_{j=0}^{RS-1} j(j-1)\ldots(j-\nu_i+1)\, w_j \psi_i^{j-\nu_i} = \sum_{j=0}^{RS-1} w_j \left[\frac{d^j}{dz^j}(z^{\nu_i} e^{\psi_i z})\right]_{z=0},$$

and thus from (10) we obtain

$$p_t = \sum_{j=0}^{RS-1} w_j \phi_j(0).$$

Now $RS \leqslant h^{2n+2}$ and so, from Lemma 5, it follows that (9) holds for all j with $0 \leqslant j \leqslant RS$. Further, by Lemma 7, we have

$$|w_j| \leqslant (8\sigma/\rho)^{RS} \leqslant (8c_{14}Lc_{15}^{L})^{RS} \leqslant c_{16}^{h^{2n+4}}.$$

Hence, since $|p_t| \geqslant 1$, we conclude that

$$0 \leqslant \log RS + c_{17}h^{2n+4} - h^{8n}.$$

The inequality is plainly impossible if h is sufficiently large and the contradiction proves the theorem.

3

LOWER BOUNDS FOR LINEAR FORMS

1. Introduction

Various conditions were obtained in Chapter 2 for the non-vanishing of the linear form

$$\Lambda = \beta_0 + \beta_1 \log \alpha_1 + \ldots + \beta_n \log \alpha_n,$$

where the α's and β's denote algebraic numbers; in particular, it suffices if $\beta_0 \neq 0$, or if $\beta_1, \ldots, \beta_n$ are linearly independent over the rationals, assuming that the α's are not 0 or 1. In the present chapter, quantitative extensions of the work will be discussed, giving positive lower bounds for $|\Lambda|$ in terms of the degrees and heights of the α's and β's; it will be recalled from Chapter 1 that the height of an algebraic number is the maximum of the absolute values of the relatively prime integer coefficients in its minimal defining polynomial. Theorems of this kind were first proved by Morduchai-Boltovskoj† in 1923, in the case $n = 1$, and by Gelfond‡ in 1935, in the case $n = 2$, $\beta_0 = 0$. A lower bound for $|\Lambda|$, valid for arbitrary n, was established in 1966, on the basis of the work described in Chapter 2, and a variety of improvements have been obtained subsequently. In particular, when the α's and also the degrees of the β's are regarded as fixed, a result that is essentially best possible has now been derived.§

Theorem 3.1. *Let $\alpha_1, \ldots, \alpha_n$ be non-zero algebraic numbers with degrees at most d and heights at most A. Further, let $\beta_0, \ldots, \beta_n$ be algebraic numbers with degrees at most d and heights at most $B\,(\geqslant 2)$. Then either $\Lambda = 0$ or $|\Lambda| > B^{-C}$, where C is an effectively computable number depending only on n, d, A and the original determinations of the logarithms.*

The estimate for C takes the form $C'(\log A)^{\kappa}$, where κ depends only on n, and C' depends only on n and d. In the case when $\beta_0 = 0$ and $\beta_1, \ldots, \beta_n$ are rational integers, it has been shown that in fact the theorem holds with $C = C'\Omega \log \Omega$, where $\Omega = (\log A)^n$; and moreover,

† *C.R.* **176** (1923), 724–7. ‡ *D.A.N.* **2** (1935), 177–82.

§ *Mat. Sbornik*, **76** (1968), 304–19; **77** (1968), 423–36 (N. I. Feldman).

if it is assumed that the height of α_j does not exceed A_j ($\geqslant 4$), then Ω can be taken as $\log A_1 \ldots \log A_n$.† Still stronger results have been obtained in the special case, of considerable importance in applications, when one of the α's, say α_n, has a large height relative to the remainder. Indeed it has been proved that if $\alpha_1, \ldots, \alpha_{n-1}$ and α_n have heights at most A' and A ($\geqslant 4$) respectively, then

$$|\Lambda| > (B \log A)^{-C \log A},$$

where $C > 0$ is effectively computable in terms of A', n and d only.‡ Further, when $\beta_0 = 0$ and $\beta_1, \ldots, \beta_n$ are rational integers, the bracketed factor $\log A$ has been eliminated to yield

$$|\Lambda| > C^{-\log A \log B},$$

which is clearly best possible with respect to A when B is fixed, and with respect to B when A is fixed.§ Furthermore, under the additional specialization $\beta_n = -1$, it has been shown that

$$|\Lambda| > A^{-C} e^{-\epsilon B}$$

for any $\epsilon > 0$, where now C depends only on A', n, d and ϵ.‖ As we shall see later, these results have particular value in connexion with the study of Diophantine problems.

It will be noted that, from the case $n = 1$ of Theorem 3.1, we have

$$|\log \alpha - \beta| > B^{-C}$$

for any algebraic number α, not 0 or 1, and for all algebraic numbers β with degrees at most d and heights at most B ($\geqslant 2$), where C depends only on d and α; more especially we have

$$|\pi - \beta| > B^{-C}$$

for some C depending only on d. Estimates of the latter kind with, in fact, precise values for C were derived long before the general result. Indeed Feldman,¶ extending work of Mahler,†† obtained the first of these inequalities with C of order $(d \log d)^2$, assuming that B is sufficiently large, and the second with C of order $d \log d$. Moreover, when β is rational, some striking inequalities of the type

$$|\pi - p/q| > q^{-42},$$

† *Acta Arith.* **27** (1974), 247–52.
‡ *Diophantine approximation and its applications* (Academic Press, 1973) pp. 1–23.
§ *Acta Arith.* **21** (1972), 117–29.
‖ *Acta Arith.* **24** (1973), 33–6
¶ *I.A.N.* **24** (1960), 357–68, 475–92.
†† *J.M.* **166** (1932), 118–50.

valid for all rationals $p/q\,(q \geqslant 2)$, were established by Mahler,† and, more recently, by similar methods, values of C arbitrarily close to the conjecturally best possible $d+1$ were derived in connexion with approximations to the logarithms of certain rational α.‡ Several further estimates of this character, classically termed transcendence measures, are furnished by the results cited after Theorem 3.1. They imply, for instance, that, subject to the hypotheses of Theorems 2.3 or 2.4, we have

$$|e^{\beta_0}\alpha_1^{\beta_1}\ldots\alpha_n^{\beta_n}-\gamma| > H^{-C\log\log H}$$

for all algebraic numbers γ with height at most $H\,(\geqslant 4)$, where C depends only on the α's, β's and the degree of γ; in particular

$$|e^{\pi}-p/q| > q^{-c\log\log q}$$

for all rationals p/q $(q \geqslant 4)$, where c denotes an absolute constant, and this is the best measure of irrationality for e^{π} obtained to date.

We shall prove here only Theorem 3.1; the demonstrations of the other results are similar, though the underlying auxiliary functions are modified, the inductive nature of the argument is more complicated, and certain lemmas appertaining to Kummer theory are employed in the latter part of the exposition in place of the determinant that occurs here. The reader is referred to the original memoirs for details. Applications of the results to various branches of number theory will be discussed in subsequent chapters.

2. Preliminaries

We begin with some observations concerning the heights of algebraic numbers. First we note that if α is an algebraic number with degree d and height H then $|\alpha| \leqslant dH$; for if α satisfies

$$a_0\alpha^d + a_1\alpha^{d-1} + \ldots + a_d = 0,$$

where the a_j denote rational integers with absolute values at most H and $a_0 \geqslant 1$, then either $|\alpha| < 1$ or

$$|\alpha| \leqslant |a_0\alpha| = |a_1 + a_2\alpha^{-1} + \ldots + a_d\alpha^{-d+1}| \leqslant dH.$$

Secondly we observe that if α, β are algebraic numbers with degrees at most d and heights at most H, then $\alpha\beta$ and $\alpha+\beta$ have degrees at most d^2 and heights at most H', where $\log H'/\log H$ is bounded above by a

† *Philos. Trans. Roy. Soc. London*, A **245** (1953), 371–98; *I.M.* **15** (1953), 30–42.
‡ *Acta Arith.* **10** (1964), 315–23.

number depending only on d. For let $\alpha^{(i)}$, $\beta^{(j)}$ denote the respective conjugates of α and β. Then $\alpha\beta$ and $\alpha+\beta$ are zeros of the polynomials

$$(ab)^{d^2}\prod_{i,j}(x-\alpha^{(i)}\beta^{(j)}),\quad (ab)^{d^2}\prod_{i,j}(x-\alpha^{(i)}-\beta^{(j)})$$

respectively, which clearly have integer coefficients and degrees at most d^2. The zeros of the minimal polynomials of $\alpha\beta$ and $\alpha+\beta$ are thus given by some subsets of the $\alpha^{(i)}\beta^{(j)}$ and $\alpha^{(i)}+\beta^{(j)}$, and the leading coefficients divide $(ab)^{d^2}$. The assertion now follows on noting that the $\alpha^{(i)}$, $\beta^{(j)}$ have absolute values at most dH.

For any integers $k \geqslant 1$, $l \geqslant 0$ we shall signify by $\nu(l;k)$ the least common multiple of $l+1, \ldots, l+k$. Further, for brevity, we shall write

$$\Delta(x;\,k) = (x+1)\ldots(x+k)/k!,$$

and we shall put $\quad \Delta(x;\,k,l,m) = \dfrac{1}{m!}\dfrac{d^m}{dx^m}(\Delta(x;\,k))^l.$

The functions have the following properties:

Lemma 1. *When x is a positive integer then also $(\nu(x;k))^m\,\Delta(x;\,k,l,m)$ is a positive integer and we have*

$$\Delta(x;\,k,l,m) \leqslant 4^{l(x+k)},\quad \nu(x;\,k) \leqslant \{c(x+k)/k\}^{2k}$$

for some absolute constant c.

Proof. First we observe that

$$\Delta(x;\,k,l,m) = (\Delta(x;\,k))^l\,\Sigma\{(x+j_1)\ldots(x+j_m)\}^{-1},$$

where the summation is over all selections of m integers $j_1, \ldots, j_m$ from the set $1, \ldots, k$ repeated l times, and the right-hand side is read as 0 if $m > kl$. Clearly $x+j_r$ divides $\nu(x;\,k)$ for each r, and since certainly $\Delta(x;\,k)$ is a rational integer, the first part of the proposition follows. Further, we see that

$$\Delta(x;\,k,l,m) \leqslant \binom{x+k}{k}^l\binom{kl}{m} \leqslant 2^{l(x+k)+kl},$$

and this gives the required estimate.

To obtain the estimate for ν, we write $\nu(x;\,k) = \nu'\nu''$, where all prime factors of ν', ν'' are $\leqslant k$ and $> k$ respectively. Since the exponent to which a prime p divides ν' is at most $\log(x+k)/\log p$, we have

$$\log\nu' \leqslant \Sigma\log(x+k) \leqslant c'k\log(x+k)/\log k,$$

where the summation is over all primes $p \leqslant k$, and c', like c, c'', c''' below, denotes an absolute constant. Now we can assume that $k > c''$ and that $x > c''k$ for some sufficiently large c'', for otherwise the desired conclusion would follow at once from the simple upper bounds $(x+k)^k$ and c^{x+k} for $\nu(x; k)$. Thus we see that

$$\nu' \leqslant \{c'''(x+k)/k\}^k.$$

But clearly ν'' divides $\Delta(x; k)$, and this does not exceed $(x+k)^k/k!$; the required estimate is now apparent. The exponent 2 can in fact be reduced easily to 1, which is best possible, but the refinement is not needed here.

We prove next a simple lemma giving a special basis for the space of polynomials with bounded degree.

Lemma 2. *If $P(x)$ is a polynomial with degree $n > 0$ and if K is a field containing its coefficients then, for any integer m with $0 \leqslant m \leqslant n$, the polynomials $P(x), P(x+1), \ldots, P(x+m)$ and $1, x, \ldots, x^{n-m-1}$ are linearly independent over K.*

Proof. The assertion is readily verified for $n = 1$. We assume the result for $n = n'$ and we proceed to prove the validity for $n = n'+1$. Suppose therefore that $0 \leqslant m \leqslant n'+1$, that $P(x)$ is a polynomial with degree $n'+1$ and that

$$R(x) = \lambda_0 P(x) + \lambda_1 P(x+1) + \ldots + \lambda_m P(x+m)$$

has degree at most $n'-m$ for some elements λ_j of K. We have

$$R(x) = (\lambda_0 + \ldots + \lambda_m)\, P(x+m+1) + \sum_{j=0}^{m} (\lambda_0 + \lambda_1 + \ldots + \lambda_j)\, Q(x+j),$$

where $Q(x) = P(x) - P(x+1)$. But $Q(x)$ has degree n' and since $P(x+m+1)$ has degree $n'+1$ we see that $\lambda_0 + \ldots + \lambda_m = 0$. It follows from the inductive hypothesis that

$$\lambda_0 + \lambda_1 + \ldots + \lambda_j = 0 \quad (0 \leqslant j \leqslant m),$$

and so $\lambda_0 = \ldots = \lambda_m = 0$, as required.

Finally we establish the non-vanishing of a particular determinant; the result will play a similar rôle to Lemma 7 of Chapter 2.

Lemma 3. *If $\omega_0, \ldots, \omega_{l-1}$ are any distinct non-zero complex numbers then the determinant of order kl with $i^r \omega_s^i$ in the $(i+1)$-th row and $(j+1)$-th column, where $j = r + sk$ $(0 \leqslant r < k,\ 0 \leqslant s < l)$, is not zero.*†

† Here $i^0 = 1$ for all i including $i = 0$.

Proof. The determinant Ω in question can plainly be expressed as a polynomial $\Omega(\omega_0, \ldots, \omega_{l-1})$ in the ω's with integer coefficients. We write

$$\Omega(z) = \Omega(z, \omega_1, \ldots, \omega_{l-1}),$$

and we observe from the Laplace expansion of Ω, taking minors formed from the first k columns, that $\Omega(z)$ is a polynomial in z with degree at most

$$\sum_{j=1}^{k} (kl-j) = k^2l - \tfrac{1}{2}k(k+1),$$

and moreover that it has a factor $z^{\frac{1}{2}k(k-1)}$. We shall prove in a moment that it also has a factor $(z-\omega_s)^{k^2}$ for each s with $1 \leqslant s < l$. This gives

$$\Omega(z) = Cz^{\frac{1}{2}k(k-1)} \prod_{s=1}^{l-1} (z-\omega_s)^{k^2},$$

where C is the coefficient of the highest power of z in $\Omega(z)$. It is easily verified that C is the product of the Vandermonde determinant of order k with typical element $(k(l-1)+i)^j$, and the determinant of order $k(l-1)$ formed like Ω, that is, with typical element $i^r\omega_s^i$, where now $1 \leqslant s < l$. The lemma follows by induction.

To prove the above proposition we begin by noting that the mth derivative $\Omega_m(z)$ of $\Omega(z)$ is given by

$$\sum (m!/(m_0! \ldots m_{k-1}!))\Omega'(m_0, \ldots, m_{k-1}, z),$$

where the summation is over all non-negative integers $m_0, \ldots, m_{k-1}$ with sum m, and $\Omega'(m_0, \ldots, m_{k-1}, z)$ is obtained from $\Omega(z)$ by replacing the element in the $(i+1)$th row and $(j+1)$th column for $j < k$ by

$$i^{r+1}(i-1)\ldots(i-m_r+1)\, z^{i-m_r}.$$

It suffices now to prove that if $m < k^2$ then the $2k$ polynomials $1, x, \ldots, x^{k-1}$ and

$$x^{r+1}(x-1)\ldots(x-m_r+1) \quad (0 \leqslant r < k)$$

are linearly dependent; for then some non-trivial linear combination of the $2k$ columns of $\Omega'(m_0, \ldots, m_{k-1}, \omega_s)$, given by

$$j < k \quad \text{and} \quad j = r+(s-1)k,$$

vanishes and so $\Omega_m(\omega_s) = 0$. To establish the linear dependence we arrange the degrees of the polynomials in ascending order, say $n_1 \leqslant n_2 \leqslant \ldots \leqslant n_{2k}$, and we observe that their sum is

$$\tfrac{1}{2}k(k-1) + \sum_{r=0}^{k-1} (r+m_r) = k(k-1)+m < 2k^2-k.$$

Thus we have $n_j < j-1$ for some j; this implies that there are j polynomials amongst the original set each with degree at most $j-2$, and these are certainly linearly dependent. The above argument clearly yields an explicit value for Ω, but only the non-vanishing is required here.

3. The auxiliary function

We come now to the proof of Theorem 3.1 and we assume accordingly that $\alpha_1, \ldots, \alpha_n$ are non-zero algebraic numbers with degrees and heights at most d and A respectively. By $C, c, c_1, c_2, \ldots$ we signify numbers, greater than 1, that depend only on n, d, A and the given determinations of the logarithms of the α's. We suppose that $\beta_0, \ldots, \beta_{n-1}$ are algebraic numbers with degrees and heights at most d and B ($\geqslant 2$) respectively such that

$$|\beta_0 + \beta_1 \log \alpha_1 + \ldots + \beta_{n-1} \log \alpha_{n-1} - \log \alpha_n| < B^{-C}, \tag{1}$$

for some sufficiently large C, and we proceed to show that there exist then rational integers $b_1', \ldots, b_n'$, not all 0, with absolute values at most c_1, satisfying

$$b_1' \log \alpha_1 + \ldots + b_n' \log \alpha_n = 0. \tag{2}$$

An inductive argument will then complete the proof of the theorem.

The subsequent work rests on the construction of an auxiliary function analogous to that obtained in Lemma 2 of Chapter 2. We signify by k an integer exceeding a sufficiently large number c as above, and we write

$$h = [\log(kB)], \quad L_{-1} = h-1, \quad L = L_0 = \ldots = L_n = [k^{1-1/(4n)}].$$

We adopt the notation of Chapter 2 with regard to partial derivatives.

Lemma 4. *There are integers $p(\lambda_{-1}, \ldots, \lambda_n)$, not all 0, with absolute values at most c_2^{hk}, such that the function*

$$\Phi(z_0, \ldots, z_{n-1}) = \sum_{\lambda_{-1}=0}^{L_{-1}} \cdots \sum_{\lambda_n=0}^{L_n} p(\lambda_{-1}, \ldots, \lambda_n)$$
$$\times (\Delta(z_0 + \lambda_{-1}; h))^{\lambda_0+1} e^{\lambda_n \beta_0 z_0} \alpha_1^{\gamma_1 z_1} \ldots \alpha_{n-1}^{\gamma_{n-1} z_{n-1}},$$

where $\gamma_r = \lambda_r + \lambda_n \beta_r$ $(1 \leqslant r < n)$, satisfies

$$|\Phi_{m_0, \ldots, m_{n-1}}(l, \ldots, l)| < B^{-\frac{1}{2}C} \tag{3}$$

for all integers l with $1 \leqslant l \leqslant h$ and all non-negative integers $m_0, \ldots, m_{n-1}$ with $m_0 + \ldots + m_{n-1} \leqslant k$.

Proof. We determine the $p(\lambda_{-1}, \ldots, \lambda_n)$ such that

$$\sum_{\lambda_{-1}=0}^{L_{-1}} \cdots \sum_{\lambda_n=0}^{L_n} p(\lambda_{-1}, \ldots, \lambda_n)\, q(\lambda_{-1}, \lambda_0, \lambda_n, l)\, \alpha_1^{\lambda_1 l} \ldots \alpha_n^{\lambda_n l} \gamma_1^{m_1} \ldots \gamma_{n-1}^{m_{n-1}} = 0 \tag{4}$$

for the above ranges of l and $m_0, \ldots, m_{n-1}$, where

$$q(\lambda_{-1}, \lambda_0, \lambda_n, z) = \sum_{\mu_0=0}^{m_0} \binom{m_0}{\mu_0} \mu_0!\, \Delta(z+\lambda_{-1}; h, \lambda_0+1, \mu_0)\, (\lambda_n \beta_0)^{m_0-\mu_0}.$$

We shall verify subsequently that (4) implies (3). Following the proof of Lemma 2 of Chapter 2, and defining the a's and b's and P' as there, we derive the same equation involving summation over $s_1, \ldots, s_n, t_0, \ldots, t_{n-1}$ as arises there, but with

$$A(s,t) = \sum_{\lambda_{-1}=0}^{L_{-1}} \cdots \sum_{\lambda_n=0}^{L_n} \sum_{\mu_0=0}^{m_0} \cdots \sum_{\mu_{n-1}=0}^{m_{n-1}} p(\lambda_{-1}, \ldots, \lambda_n)\, q' q'' q''',$$

where now

$$q''' = \binom{m_0}{\mu_0} \mu_0!\, \Delta(l+\lambda_{-1}; h, \lambda_0+1, \mu_0)\, \lambda_n^{m_0-\mu_0} b_n^{\mu_0} b_{0,t_0}^{(m_0-\mu_0)}$$

and the $b_{rt}^{(j)}$ have absolute values at most $(2B)^j$. Thus we conclude that (4) will be satisfied if the d^{2n} equations $A(s,t) = 0$ hold. Now these represent $M \leqslant d^{2n}h(k+1)^n$ linear equations in the

$$N = (L_{-1}+1) \ldots (L_n+1)$$

unknowns $p(\lambda_{-1}, \ldots, \lambda_n)$. Further, Lemma 1 shows that, after multiplying by $(\nu(0; 3h))^{m_0}$, the coefficients in these equations will be rational integers. Furthermore we have

$$\Delta(l+\lambda_{-1}; h, \lambda_0+1, \mu_0) \leqslant c_3^{Lh},$$

and, since $kB \leqslant e^{h+1}$, we see that

$$|q'| \leqslant c_4^{Lh}, \quad |q''| \leqslant e^{2h(m_1+\ldots+m_{n-1})},$$

$$|q'''| \leqslant 2^{m_0} (\mu_0 b_n)^{\mu_0} (2B\lambda_n)^{m_0-\mu_0} c_3^{Lh} \leqslant e^{2hm_0} c_3^{Lh}.$$

Since also $\nu(0; 3h) \leqslant c_5^h$, it follows that the coefficients have absolute values at most $U = c_6^{hk}$. Now $N > hk^{n+\frac{1}{2}} > 2M$ and hence, by Lemma 1 of Chapter 2, the system of equations $A(s,t) = 0$ can be solved non-trivially and the integers $p(\lambda_{-1}, \ldots, \lambda_n)$ can be chosen to have absolute values at most $NU \leqslant c_2^{hk}$.

It remains only to verify that (4) implies (3). Now the left-hand side of (4) is obtained from the number on the left of (3), omitting modulus signs and a factor

$$P = (\log \alpha_1)^{m_1} \ldots (\log \alpha_{n-1})^{m_{n-1}},$$

by substituting α_n for $\alpha'_n = e^{\beta_0}\alpha_1^{\beta_1}\ldots\alpha_{n-1}^{\beta_{n-1}}$. From (1) we have

$$|\log\alpha'_n - \log\alpha_n| < B^{-C},$$

for some value of the first logarithm and since, for any complex number z, $|e^z - 1| \leqslant |z|\,e^{|z|}$, we obtain

$$|\alpha'_n - \alpha_n| < B^{-\frac{1}{2}C}. \tag{5}$$

Also we have
$$|\alpha_n'^{\lambda_n l} - \alpha_n^{\lambda_n l}| \leqslant c_7^{Ll}|\alpha'_n - \alpha_n|,$$

and estimates similar to those employed above show that

$$|P| \leqslant c_8^k, \quad |q(\lambda_{-1}, \lambda_0, \lambda_n, l)| \leqslant c_9^{(L+m_0)h}, \quad |\gamma_r| \leqslant e^{2h}, \quad |\alpha_r^{\lambda_r l}| \leqslant c_{10}^{Lh}.$$

Thus we see that the number on the left of (3) is at most $Nc_{11}^{hk}B^{-\frac{1}{2}C}$. But clearly $N \leqslant e^{2hn}$ and $h \leqslant \log(kB)$, and hence (3) follows if $C > c_{12}k\log k$.

Lemma 5. *Let $m_0, \ldots, m_{n-1}$ be any non-negative integers with*

$$m_0 + \ldots + m_{n-1} \leqslant k,$$

and let
$$f(z) = \Phi_{m_0,\ldots,m_{n-1}}(z, \ldots, z).$$

Then, for any number z, we have $|f(z)| \leqslant c_{13}^{hk+L|z|}$. Further, for any integer l with $h < l \leqslant hk^{8n}$, either $|f(l)| < B^{-\frac{1}{4}C}$ or

$$|f(l)| > c_{14}^{-hk(1+\log(l/h))-Ll}. \tag{6}$$

Proof. The function $f(z)$ is given by

$$P\sum_{\lambda_{-1}=0}^{L_{-1}}\cdots\sum_{\lambda_n=0}^{L_n} p(\lambda_{-1}, \ldots, \lambda_n)\, q(\lambda_{-1}, \lambda_0, \lambda_n, z)$$
$$\times\, e^{\lambda_n\beta_0 z}\alpha_1^{\gamma_1 z}\ldots\alpha_{n-1}^{\gamma_{n-1}z}\gamma_1^{m_1}\ldots\gamma_{n-1}^{m_{n-1}},$$

where P and $q(\lambda_{-1}, \lambda_0, \lambda_n, z)$ are defined as in Lemma 4. Now (5) implies that $|\alpha_n'^{z}| \leqslant c_{15}^{|z|}$ and clearly

$$|\alpha_1^{\lambda_1 z}\ldots\alpha_{n-1}^{\lambda_{n-1}z}| \leqslant c_{16}^{L|z|}.$$

Furthermore, since
$$|z + \lambda_{-1}| \leqslant [|z|] + h,$$

we deduce from Lemma 1 that

$$|\Delta(z + \lambda_{-1};\, h, \lambda_0 + 1, \mu_0)| \leqslant c_{17}^{L(|z|+h)}.$$

This gives
$$|q(\lambda_{-1}, \lambda_0, \lambda_n, z)| \leqslant e^{2hm_0}c_{17}^{L(|z|+h)},$$

and the required estimate now follows easily as in the latter part of the proof of Lemma 4.

To prove the second assertion, we begin by noting that the expression on the left of (4), say Q, is an algebraic number with degree at most d^{2n}. Further, by estimates similar to those given above, it is readily verified that each conjugate of Q, obtained by substituting arbitrary conjugates for the α's and β's, has absolute value at most c_{18}^{hk+Ll}. Furthermore, from Lemma 1, we see that on multiplying Q by

$$(\nu(l;2h))^{m_0} P' \leqslant (c_{19}\, l/h)^{4hm_0} c_{20}^{hk+Ll},$$

one obtains an algebraic integer. Hence we conclude that either $Q = 0$ or

$$|Q| \geqslant c_{21}^{-hk-Ll}(l/h)^{-c_{22}hm_0}.$$

Since $m_0 \leqslant k$, the number on the right of the last inequality exceeds the right-hand side of (6) for some c_{14}. Further, as above, we deduce easily from (5) that $P^{-1}f(l)$ differs from Q by at most $c_{23}^{lk} B^{-\frac{1}{8}C}$. But if $l \leqslant hk^{8n}$ and $C > k^{8n+2}$, then this is at most $\frac{1}{2}|Q|$, and hence, if $Q \neq 0$, we obtain $|f(l)| > \frac{1}{2}|PQ|$, which gives (6).

Lemma 6. *Suppose that* $0 < \epsilon < c^{-1}$ *for some sufficiently large* c. *Then, for any integer* J *with* $0 \leqslant J < 8n/\epsilon$, (3) *is satisfied for all integers* l *with* $1 \leqslant l \leqslant hk^{\epsilon J}$ *and all non-negative integers* $m_0, \ldots, m_{n-1}$ *with* $m_0 + \ldots + m_{n-1} \leqslant k/2^J$.

Proof. The lemma holds for $J = 0$ by Lemma 4. We suppose that K is an integer with $0 \leqslant K \leqslant (8n/\epsilon) - 1$ and we assume that the lemma has been verified for $J = 0, 1, \ldots, K$. We proceed to prove the proposition for $J = K+1$.

It suffices to show that for any integer l with $R_K < l \leqslant R_{K+1}$ and any set of non-negative integers $m_0, \ldots, m_{n-1}$ with $m_0 + \ldots + m_{n-1} \leqslant S_{K+1}$, we have $|f(l)| < B^{-\frac{1}{8}C}$, where $f(z)$ is defined as in Lemma 5, and

$$R_J = [hk^{\epsilon J}], \quad S_J = [k/2^J] \quad (J = 0, 1, \ldots).$$

From our inductive hypothesis we deduce, as in Lemma 4 of Chapter 2, that

$$|f_m(r)| < n^k B^{-\frac{1}{8}C} \quad (1 \leqslant r \leqslant R_K,\ 0 \leqslant m \leqslant S_{K+1}). \tag{7}$$

We write, for brevity,

$$F(z) = \{(z-1)\ldots(z-R_K)\}^{S+1},$$

where $S = S_{K+1}$, and we denote by Γ the circle in the complex plane, described in the positive sense, with centre the origin and radius $R = R_{K+1}k^{1/(8n)}$. By Cauchy's residue theorem we have

$$\frac{1}{2\pi i}\int_{\Gamma} \frac{f(z)\,dz}{(z-l)\,F(z)} = \frac{f(l)}{F(l)} + \frac{1}{2\pi i}\sum_{r=1}^{R_K}\sum_{m=0}^{S}\frac{f_m(r)}{m!}\int_{\Gamma_r}\frac{(z-r)^m\,dz}{(z-l)\,F(z)}, \tag{8}$$

where Γ_r denotes the circle in the complex plane, described in the positive sense, with centre r and radius $\frac{1}{2}$; for the residue of the pole of the integrand on the left at $z = r$ is given by

$$\frac{1}{S!}\frac{d^S}{dz^S}\left\{\frac{(z-r)^{S+1}f(z)}{(z-l)\,F(z)}\right\},$$

evaluated at $z = r$, and the integral over Γ_r on the right is given by

$$\frac{2\pi i}{(S-m)!}\frac{d^{S-m}}{dz^{S-m}}\left\{\frac{(z-r)^{S+1}}{(z-l)\,F(z)}\right\},$$

again evaluated at $z = r$, and (8) now follows by Leibnitz's theorem. Since, for z on Γ_r,

$$|(z-r)^m/F(z)| \leqslant \{\tfrac{1}{8}(R_K-r-1)!\,(r-2)!\}^{-S-1} \leqslant 8^{R_K S}(R_K!)^{-S-1},$$

we deduce from (7) that the absolute value of the double sum on the right of (8) is at most

$$R_K(S+1)\,8^{R_K S+1}\,(R_K!)^{-S-1}\,B^{-\frac{1}{4}C}.$$

Further, for $R_K < l \leqslant R_{K+1}$, we have

$$|F(l)| = \{(l-1)!/(l-R_K-1)!\}^{S+1} \leqslant (2^{R_{K+1}}R_K!)^{S+1},$$

and, since $R_{K+1} \leqslant hk^{8n}$, we see that if (6) holds then $|f(l)| > B^{-\frac{1}{4}C}$, whence the number on the right of (8) exceeds $\frac{1}{2}|f(l)/F(l)|$. We proceed to show that the assumption that (6) is valid leads to a contradiction.

Let θ and Θ denote respectively the upper bound of $|f(z)|$ and the lower bound of $|F(z)|$ with z on Γ. Since $2|z-l|$ with z on Γ exceeds the the radius of Γ, we obtain from (8)

$$4\theta\,|F(l)| > \Theta\,|f(l)|. \tag{9}$$

Now clearly we have $\Theta \geqslant (\frac{1}{2}R)^{R_K(S+1)}$ and thus

$$\log(\Theta\,|F(l)|^{-1}) \geqslant R_K(S+1)\log(\tfrac{1}{2}k^{1/(8n)}). \tag{10}$$

Further, from Lemma 5, we see that $\theta \leqslant c_{13}^{hk+LR}$ and so, by virtue of (6),

$$\log(\theta\,|f(l)|^{-1}) \leqslant c_{25}\{LR+hk\log(R_{K+1}/h)\}. \tag{11}$$

But the number on the right of (10) is at least

$$2^{-K-6}n^{-1}hk^{\epsilon K+1}\log k,$$

and that on the right of (11) is at most

$$c_{25}\,hk\{\epsilon(K+1)\log k + k^{\epsilon(K+1)-1/(8n)}\}.$$

If $\epsilon^{-1} > c > 2^7 n c_{25}$ and k is sufficiently large, the estimates are plainly inconsistent. The contradiction implies the validity of (3) and the lemma follows by induction.

Lemma 7. *For all integers l with $0 \leqslant l \leqslant hk^{4n}$ we have*

$$\sum_{\lambda_{-1}=0}^{L_{-1}} \cdots \sum_{\lambda_n=0}^{L_n} p(\lambda_{-1}, \ldots, \lambda_n)\,(\Delta(\lambda_{-1}+l/k;\, h))^{\lambda_0+1}\,\alpha_1^{\lambda_1 l/k} \ldots \alpha_n^{\lambda_n l/k} = 0. \tag{12}$$

Proof. From Lemma 6 we see that (3) holds for all integers l with $1 \leqslant l \leqslant X$ and all non-negative integers $m_0, \ldots, m_{n-1}$ with

$$m_0 + \ldots + m_{n-1} \leqslant Y,$$

where
$$X = [hk^{7n}], \quad Y = [c_{26}^{-1} k],$$

and $c_{26} = 2^{8n/\epsilon}$. It follows as in the proof of Lemma 6 that

$$f(z) = \Phi(z, \ldots, z)$$

satisfies
$$|f_m(r)| < n^k B^{-\frac{1}{8}C} \quad (1 \leqslant r \leqslant X,\ 0 \leqslant m \leqslant Y). \tag{13}$$

Now let l be any integer with $0 \leqslant l \leqslant hk^{4n}$ and define

$$E(z) = \{(z-1)\ldots(z-X)\}^{Y+1},$$

with the proviso that the factor $(z-l/k)$ is excluded if l/k is an integer. Denoting by Γ the circle in the complex plane, described in the positive sense, with centre the origin and radius $R = Xk^{1/(8n)}$, we deduce from Cauchy's residue theorem

$$\frac{1}{2\pi i}\int_\Gamma \frac{f(z)\,dz}{(z-l/k)\,E(z)} = \frac{f(l/k)}{E(l/k)} + \frac{1}{2\pi i}\sum_{r=1}^{X}{}' \sum_{m=0}^{Y} \frac{f_m(r)}{m!}\int_{\Gamma_r} \frac{(z-r)^m\,dz}{(z-l/k)\,E(z)},$$

where the dash signifies that $r = l/k$, if an integer, is excluded from the summation, and Γ_r denotes the circle in the complex plane, described in the positive sense, with centre r and radius $1/(2k)$. Since, for z on Γ_r,

$$|(z-r)^m/E(z)| \leqslant \{(8kX)^{-1}(X-r-1)!\,(r-2)!\}^{-Y-1} \leqslant 8^{3XY}(X!)^{-Y-1},$$

it follows from (13) that the absolute value of the double sum on the right of the above equation is at most

$$X(Y+1)\,8^{3XY}(X!)^{-Y-1}B^{-\frac{1}{8}C}.$$

Further, by virtue of Lemma 5, we have, for any z on Γ, $|f(z)| < c_{13}^{hk+LR}$, and it is clear that

$$|E(z)| \geqslant (\tfrac{1}{2}R)^{(X-1)(Y+1)},$$

$$|E(l/k)| \leqslant (2X)^{X(Y+1)} \leqslant 8^{X(Y+1)}(X!)^{Y+1}.$$

Thus we obtain

$$|f(l/k)| < c_{13}^{hk+LR}(8^{-3}k^{1/(8n)})^{-XY} + B^{-\frac{1}{8}C},$$

and, since $Lk^{1/(8n)} < k$, we deduce easily that the number on the right is at most e^{-XY}.

Now clearly the left-hand side of (12), say Q, is an algebraic number with degree at most $(dk)^n$, and each conjugate has absolute value at most $c_{27}^{hk^{4n}}$. Further, it is readily verified that on multiplying Q by

$$(a_1 \dots a_n)^{Ll} k^{2h(L+1)} \leqslant c_{28}^{hk^{4n+1}}$$

one obtains an algebraic integer; for certainly the denominator of either $k^h/h!$ or $\Delta(\lambda_{-1}+l/k;\, h)$, expressed in lowest terms, is free of a given prime p according as p does or does not divide k. Thus, if $Q \neq 0$, we have $|Q| > c_{29}^{-hk^{6n}}$. But it is easily seen from (5) that

$$|Q - f(l/k)| < B^{-\frac{1}{8}C},$$

whence $|f(l/k)| > \frac{1}{2}|Q|$. The estimate for $|Q|$ given above is plainly inconsistent with the upper bound e^{-XY} for $|f(l/k)|$ obtained earlier, and thus we conclude that $Q = 0$, as required.

4. Proof of main theorem

First we observe that, by virtue of Lemma 2, the polynomials

$$(\Delta(\lambda_{-1}+x;\, h))^{\lambda_0+1} \quad (0 \leqslant \lambda_{-1} \leqslant L_{-1},\ 0 \leqslant \lambda_0 \leqslant L_0)$$

are linearly independent over the rationals. Thus, on writing

$$\sum_{\lambda_{-1}=0}^{L_{-1}} \sum_{\lambda_0=0}^{L_0} p(\lambda_{-1}, \dots, \lambda_n)\,(\Delta(\lambda_{-1}+x;\, h))^{\lambda_0+1} = \sum_{\lambda'=0}^{L'} p'(\lambda', \lambda_1, \dots, \lambda_n)\, x^{\lambda'},$$

where $L' = h(L+1)$, we see that one at least of the $L'' = (L'+1)(L+1)^n$ numbers $p'(\lambda', \lambda_1, \dots, \lambda_n)$ is not 0. Now (12) can be written in the form

$$\sum_{\lambda'=0}^{L'} \sum_{\lambda_1=0}^{L_1} \dots \sum_{\lambda_n=0}^{L_n} p'(\lambda', \lambda_1, \dots, \lambda_n)\,(l/k)^{\lambda'} (\alpha_1^{\lambda_1/k} \dots \alpha_n^{\lambda_n/k})^l = 0,$$

and, by Lemma 7, the equation holds in particular for $0 \leqslant l \leqslant L''$. It follows that the determinant of order L'', given by the terms involving l only, vanishes. But the determinant is of the kind indicated in Lemma 3, and thus we conclude that

$$\alpha_1^{\lambda_1/k} \dots \alpha_n^{\lambda_n/k} = \alpha_1^{\lambda_1'/k} \dots \alpha_n^{\lambda_n'/k}$$

for some distinct sets $\lambda_1, \ldots, \lambda_n$ and $\lambda'_1, \ldots, \lambda'_n$. This gives

$$b'_1 \log \alpha_1 + \ldots + b'_n \log \alpha_n = (2\pi i) jk$$

for some rational integer j, where $b'_r = \lambda_r - \lambda'_r$. Clearly we have $|b'_r| \leqslant 2L$, and since $L \leqslant k^{1-1/(4n)}$ it follows that the number on the left has absolute value less than $2\pi k$. Hence we conclude that $j = 0$, and so (2) holds, as required.

The proof of the theorem is now completed by induction. Suppose that $\beta_0, \ldots, \beta_n$ are given as in the enunciation and that $0 < |\Lambda| < B^{-2C}$ for some sufficiently large C. Then one at least of $\beta_1, \ldots, \beta_n$ is not 0, and we shall assume that in fact $\beta_n \neq 0$. By the preliminary observations in §2, we see that (1) holds with β_j $(1 \leqslant j < n)$ replaced by $\beta'_j = -\beta_j/\beta_n$ and further that the β'_j have degrees at most d^2 and heights at most $B' \leqslant B^c$ for some c depending only on d. Hence we conclude that (2) holds for some $b'_1, \ldots, b'_n$ as indicated in § 3. Now if $b'_r \neq 0$ we have

$$0 < |\Lambda'| < c_1 B^{-C},$$

where Λ' is obtained from Λ by replacing β_j with

$$\beta''_j = b'_r \beta_j - b'_j \beta_r \quad (0 \leqslant j < n),$$

b'_0 being defined as 0. Further, the observations in § 2 show that β''_j has degree at most d^2 and height at most $B'' \leqslant B^c$ for some $c = c(n, d, A)$. Furthermore we have $\beta''_r = 0$. But the theorem is plainly valid for $n = 0$, and if we assume that it holds for fewer than n logarithms then the above shows that it will also hold for n logarithms. This establishes the result.

It will be noted that the inductive argument would not be needed if $\log \alpha_1, \ldots, \log \alpha_n$ were linearly independent over the rationals, and moreover Lemma 7 would not be required if $\alpha_1, \ldots, \alpha_n$ were multiplicatively independent.

4

DIOPHANTINE EQUATIONS

1. Introduction

Diophantine analysis pertains, in general terms, to the study of the solubility of equations in integers. Although researches in this field have their roots in antiquity and a history of the subject amounts, more or less, to a history of mathematics itself, it is only in relatively recent times that there have emerged any general theories, and we shall accordingly begin our discussion in 1900 by referring again to Hilbert's famous list of problems.

The tenth of these asked for a universal algorithm for deciding whether or not a given Diophantine equation, that is, an equation $f(x_1, \ldots, x_n) = 0$, where f denotes a polynomial with integer coefficients, is soluble in integers $x_1, \ldots, x_n$. Though Hilbert posed his question in terms of solubility, there are, of course, many other sorts of information that one might like to have in this connexion; for instance, one might enquire as to whether a particular equation has infinitely many solutions, or one might seek some description of the distribution or size of the solutions. In 1970, Matijasevic,† developing work of Davis, Robinson and Putnam,‡ proved that a general algorithm of the type sought by Hilbert does not in fact exist. A more realistic problem arises, however, if one limits the number of variables, and for, in particular, polynomials in two unknowns our knowledge is now quite substantial.

A full account of the early results in this field is furnished by Dickson's celebrated *History of the theory of numbers*; here references are given to a diverse multitude of Diophantine problems that were investigated by a wide variety of *ad hoc* methods mainly during the last two centuries. The first major advance towards a coherent theory was made by Thue§ in 1909 when he proved that the equation $F(x, y) = m$, where F denotes an irreducible binary form with integer coefficients and degree at least 3, possesses only a finite number of solutions in integers x, y. Thue established the result by way of his

† *D.A.N.* **191** (1970), 279–82. ‡ *Ann. Math.* **74** (1961), 425–36.
§ *J.M.* **135** (1909), 284–305.

fundamental studies on rational approximations to algebraic numbers; on writing the equation in the form

$$a(x-\alpha_1 y)\dots(x-\alpha_n y) = m,$$

one sees that one of the zeros α of $F(x, 1)$ has a rational approximation x/y $(y > 0)$ with $|\alpha - x/y| < c/y^n$ for some c depending only on F and m, and Thue showed that this is impossible if y is sufficiently large.† Thue's work was much extended by Siegel‡ in 1929; Siegel proved that the equation $f(x, y) = 0$, where f denotes a polynomial with integer coefficients, has only a finite number of solutions in integers x, y if the curve it represents has genus 1 or genus 0 and at least three infinite valuations; otherwise the curve can be parameterized and there are then infinitely many so-called 'ganzartige' solutions, that is, algebraic solutions with constant denominators. Siegel's work depended upon, amongst other things, an improved version of Thue's approximation result which he obtained in 1921,§ and the famous Mordell–Weil theorem,‖ proved in 1928, on the finiteness of the basis of the group of rational points on the curve. The work of Thue and Siegel satisfactorily settles the question as to which curves possess only finitely many integer points and, moreover, it yields an estimate for the number of points when finite. But it throws no light on the basic Hilbert problem as to whether or not such points exist and, even less therefore, does it provide an algorithm for determining their totality; for the arguments depend on an assumption, made at the outset, that the equation has at least one large solution and this is purely hypothetical. Another proof of Thue's theorem, under a mild restriction, was given by Skolem¶ in 1935 by means of a p-adic argument very different from the original, but here the work depends on the compactness property of the p-adic integers and so is again non-effective.

Our purpose here is to apply the work of Chapter 3 to effectively resolve a wide class of Diophantine equations. In particular we shall treat the Thue equation $F(x, y) = m$ defined over any algebraic number field, the famous Mordell equation $y^2 = x^3 + k$, to which, incidentally, there attaches a history dating back to Bachet in 1621, and we shall obtain an effective algorithm for determining all the integer points on an arbitrary curve of genus 1. Our theorems will be proved in an essentially qualitative form, but it will be apparent that

† See Chapter 7.

‡ *Abh. Preuss. Akad. Wiss.* (1929), no. 1.

§ *M.Z.* **10** (1921), 173–213.

‖ *Acta Math.* **53** (1928), 281–315.

¶ *M.A.* **111** (1935), 399–424.

they can be adapted to yield explicit bounds for the sizes of all the solutions of the equations. A summary of quantitative aspects of the work is given in the last section.

2. The Thue equation

Let K be an algebraic number field with degree d, let $\alpha_1, \ldots, \alpha_n$ be $n \geqslant 3$ distinct algebraic integers in K, and let μ be any non-zero algebraic integer in K. We prove:

Theorem 4.1. *The equation*

$$(X - \alpha_1 Y) \ldots (X - \alpha_n Y) = \mu$$

has only a finite number of solutions in algebraic integers X, Y in K and these can be effectively determined.

We define the *size* of any algebraic integer θ in K as the maximum of the absolute values of its conjugates, and we signify the size of θ by $\|\theta\|$. With this notation, we shall in fact show how one can obtain an explicit bound for $\|X\|$ and $\|Y\|$ for all X, Y as above. The bound can be expressed in terms of d and the maximum of the heights of $\alpha_1, \ldots, \alpha_n, \mu$ and some algebraic integer generating K; we shall denote by $c_1, c_2, \ldots$ positive numbers that can be specified in terms of these quantities only. We shall assume that K has s conjugate real fields and $2t$ conjugate complex fields so that $d = s + 2t$; further we shall signify by $\theta^{(1)}, \ldots, \theta^{(d)}$ the conjugates of any element θ of K, with $\theta^{(1)}, \ldots, \theta^{(s)}$ real and $\theta^{(s+1)}, \ldots, \theta^{(s+t)}$ the complex conjugates of $\theta^{(s+t+1)}, \ldots, \theta^{(d)}$ respectively. The subsequent arguments rest on the well-known result, dating back to Dirichlet, that there exist $r = s + t - 1$ units $\eta_1, \ldots, \eta_r$ in K such that

$$|\log|\eta_i^{(j)}|| < c_1 \quad (1 \leqslant i, j \leqslant r)$$

and $|\Delta| > c_2$, where Δ denotes the determinant of order r with $\log|\eta_i^{(j)}|$ in the ith row and jth column.†

We suppose now that X, Y are any algebraic integers in K satisfying the given equation and we write, for brevity,

$$\beta_i = X - \alpha_i Y \quad (1 \leqslant i \leqslant n).$$

We denote by $N\beta_i$ the field norm of β_i and we put $m_i = |N\beta_i|$, so that $m_1 \ldots m_n = |N\mu|$. We proceed first to show that an associate γ_i of β_i

† Cf. Hecke (Bibliography).

can be determined such that

$$|\log|\gamma_i^{(j)}|| < c_3 \quad (1 \leqslant j \leqslant d). \tag{1}$$

This follows in fact from the observation that every point P in r-dimensional Euclidean space occurs within a distance c_4 of some point of the lattice with basis

$$(\log|\eta_i^{(1)}|, \ldots, \log|\eta_i^{(r)}|) \quad (1 \leqslant i \leqslant r).$$

On taking P as the point

$$(\log|\beta_i^{(1)}|, \ldots, \log|\beta_i^{(r)}|),$$

we deduce that there exist rational integers $b_{i1}, \ldots, b_{ir}$ such that

$$\gamma_i = \beta_i \eta_1^{b_{i1}} \ldots \eta_r^{b_{ir}} \tag{2}$$

satisfies (1) for $1 \leqslant j \leqslant r$, with c_4 in place of c_3, and since

$$|\gamma_i^{(j+t)}| = |\gamma_i^{(j)}| \quad (s < j \leqslant s+t),$$

the same holds for $1 \leqslant j \leqslant d$ except possibly for $j = s+t$ and $j = s+2t$ (only one of which exists if $t = 0$). But we have

$$|\gamma_i^{(1)} \ldots \gamma_i^{(d)}| = m_i, \quad 1 \leqslant m_i \leqslant |N\mu| \leqslant c_5,$$

whence (1) holds for all j, as required.

Now let $H_i = \max|b_{ij}|$ and let l be a suffix for which $H_l = \max H_i$. The equations

$$\log|\gamma_i^{(j)}/\beta_i^{(j)}| = b_{i1}\log|\eta_1^{(j)}| + \ldots + b_{ir}\log|\eta_r^{(j)}| \quad (1 \leqslant j \leqslant r)$$

serve to express each number Δb_{ij} as a linear combination of the numbers on the left with coefficients given by cofactors of Δ, and thus the maximum of the absolute values of these numbers exceeds $c_6 H_i$. Let the maximum be given by $j = J$. Then from (1) we have

$$|\log|\beta_i^{(J)}|| = |\log|\beta_i^{(J)}/\gamma_i^{(J)}| + \log|\gamma_i^{(J)}|| > c_6 H_i - c_3,$$

and, since $|\beta_i^{(1)} \ldots \beta_i^{(d)}| = m_i$, it follows that

$$\log|\beta_i^{(h_i)}| < -(c_6 H_i - c_3 - \log m_i)/(d-1)$$

for some h_i. Thus, if $H_l > c_7$, we have $|\beta_l^{(h)}| < e^{-c_8 H_l}$ for some h. Further, since

$$\beta_1^{(h)} \ldots \beta_n^{(h)} = \mu^{(h)},$$

we obtain $|\beta_k^{(h)}| > c_9$ for some $k \neq l$. We take j to be any suffix other than k or l; this exists since, by hypothesis, $n \geqslant 3$.

We may now, for simplicity, omit the superscript h and assume that $\alpha_i^{(h)} = \alpha_i$, $\beta_i^{(h)} = \beta_i$. From the identity

$$(\alpha_k - \alpha_l)\beta_j + (\alpha_l - \alpha_j)\beta_k + (\alpha_j - \alpha_k)\beta_l = 0,$$

we obtain

$$\eta_1^{b_1} \dots \eta_r^{b_r} - \alpha = \omega,$$

where

$$b_s = b_{ks} - b_{js} \quad (1 \leqslant s \leqslant r),$$

$$\alpha = \frac{(\alpha_j - \alpha_l)\gamma_k}{(\alpha_k - \alpha_l)\gamma_j}, \quad \omega = \frac{(\alpha_k - \alpha_j)\beta_l\gamma_k}{(\alpha_k - \alpha_l)\beta_k\gamma_j}.$$

On noting that, for any complex number z, the inequality $|e^z - 1| < \frac{1}{4}$ implies that

$$|z - ib\pi| \leqslant 4\,|e^z - 1|,$$

for some rational integer b, we deduce easily, on taking

$$z = b_1 \log \eta_1 + \dots + b_r \log \eta_r - \log \alpha,$$

where the logarithms have their principal values, that, if $|\omega/\alpha| < \frac{1}{4}$, then $|\Lambda| \leqslant 4\,|\omega/\alpha|$, where $\Lambda = z - b \log(-1)$. Clearly $\omega \neq 0$ and so also $\Lambda \neq 0$. Further we see that $|b_j| \leqslant 2H_l$ for all j, and so the imaginary part of z has absolute value at most πB, where $B = 4rH_l$. Thus we have $|b| \leqslant B$, and certainly $|b_j| \leqslant B$. Furthermore, from the estimates for $\beta_k = \beta_k^{(h)}$ and $\beta_l = \beta_l^{(h)}$ given above, we see that, if $H_l > c_{10}$, then

$$4|\omega/\alpha| < c_{11}|\beta_l/\beta_k| < e^{-c_{12}B}.$$

But $\eta_1, \dots, \eta_r$ and α have degrees at most d, and their heights are bounded above by a number c_{13}. Hence Theorem 3.1 gives $|\Lambda| > B^{-C}$ for some C as above, and from this and our estimate $|\Lambda| < e^{-c_{12}B}$ we conclude that $B < c_{14}$, whence $H_l < c_{15}$. It follows from (1) and (2) that

$$\|\beta_i\| < e^{c_{16}H_l} < c_{17},$$

and now the equations

$$X = \frac{\alpha_2\beta_1 - \alpha_1\beta_2}{\alpha_2 - \alpha_1}, \quad Y = \frac{\beta_1 - \beta_2}{\alpha_2 - \alpha_1}$$

and their conjugates clearly imply the validity of Theorem 4.1.

3. The hyperelliptic equation

As in § 2, we signify by K an algebraic number field with degree d. We suppose that $\alpha_1, \dots, \alpha_n$ are $n \geqslant 3$ algebraic integers in K with, say, $\alpha_1, \alpha_2, \alpha_3$ distinct and different from $\alpha_4, \dots, \alpha_n$, and we prove:

Theorem 4.2. *The equation*

$$Y^2 = (X-\alpha_1)\dots(X-\alpha_n) \tag{3}$$

has only a finite number of solutions in algebraic integers X, Y in K and these can be effectively determined.

We shall establish Theorem 4.2 from Theorem 4.1 by a method of Siegel,† and again it will be clear that the arguments enable one to furnish explicit bounds for $\|X\|$ and $\|Y\|$. The conclusion of Theorem 4.2 plainly remains valid if a non-zero factor in K is introduced on the right of (3), and thus the theorem covers, in particular, the elliptic equation

$$y^2 = ax^3+bx^2+cx+d,$$

where all quantities signify rational integers. In this case, however, the result can be derived from Theorem 4.1 by a readier method, due to Mordell, involving the theory of the reduction of binary quartic forms.‡

Suppose now that X, Y are non-zero algebraic integers in K satisfying (3). We show first that there exist algebraic integers ξ_j, η_j, ζ_j $(j = 1,2,3)$ in K with

$$X-\alpha_j = (\xi_j/\eta_j)\,\zeta_j^2, \tag{4}$$

$$\max(\|\xi_j\|, \|\eta_j\|) < c_1,$$

where c_1, like $c_2, c_3, \dots$, denotes a positive number specified as in § 2, that is, depending only on d and the maximum of the heights of $\alpha_1, \dots, \alpha_n$ and some algebraic integer generating K. For simplicity we write $\alpha = \alpha_j$, and we observe that, by virtue of the ideal equation

$$[Y^2] = [X-\alpha_1]\dots[X-\alpha_n],$$

we have

$$[X-\alpha] = \mathfrak{a}\mathfrak{b}^2$$

for some ideals $\mathfrak{a}, \mathfrak{b}$ in K, where $\mathfrak{a}$ divides

$$\prod_{i \neq j} [\alpha-\alpha_i].$$

Further, there exist ideals $\mathfrak{a}'$, $\mathfrak{b}'$ in the ideal classes inverse to those of $\mathfrak{a}, \mathfrak{b}$ respectively with norms at most c_2, and clearly $\mathfrak{a}\mathfrak{a}'$ and $\mathfrak{a}'\mathfrak{b}'^2$ are principal ideals; the latter are therefore generated by algebraic integers ξ', η' in K with

$$|N\xi'| \leqslant c_2 N\mathfrak{a}, \quad |N\eta'| \leqslant c_2^3.$$

† *J. London Math. Soc.* **1** (1926), 66–8.
‡ *J. London Math. Soc.* **43** (1968), 1–9.

Furthermore, since $$N\mathfrak{a} \leqslant \prod_{i \neq j} N[\alpha - \alpha_j] < c_3,$$

it follows easily, as in the derivation of (1), that there exist associates ξ'', η'' of ξ', η' respectively satisfying

$$\max(\|\xi''\|, \|\eta''\|) < c_4.$$

Now $\mathfrak{b}\mathfrak{b}'$ is principal and is therefore generated by some algebraic integer ζ' in K. Hence from the equation

$$(\mathfrak{a}'\mathfrak{b}'^2)[X-\alpha] = (\mathfrak{a}\mathfrak{a}')(\mathfrak{b}\mathfrak{b}')^2$$

we obtain $$X - \alpha = \epsilon(\xi''/\eta'')\zeta'^2,$$

where ϵ denotes a unit in K. By Dirichlet's theorem there exists a fundamental system $\epsilon_1, \ldots, \epsilon_r$ of units in K satisfying

$$\max(\|\epsilon_1\|, \ldots, \|\epsilon_r\|) < c_5,$$

and we have $$\epsilon = \rho\epsilon_1^{j_1} \ldots \epsilon_r^{j_r}$$

for some rational integers $j_1, \ldots, j_r$ and some root of unity ρ; it is now clear that the numbers ξ, η, ζ given by

$$\xi''\rho\epsilon_1^{j_1'} \ldots \epsilon_r^{j_r'}, \quad \eta'', \quad \zeta'\epsilon_1^{\frac{1}{2}(j_1-j_1')} \ldots \epsilon_r^{\frac{1}{2}(j_r-j_r')}$$

respectively, where $j_i' = 0$ or 1 according as j_i is even or odd, have the required properties.

On eliminating X from (4) we obtain three equations of the form

$$\sigma_2\zeta_2^2 - \sigma_3\zeta_3^2 = \alpha_3 - \alpha_2,$$

where $\sigma_j = \xi_j/\eta_j$ $(j = 1, 2, 3)$. Further, on writing

$$\beta_1 = \sigma_2^{\frac{1}{2}}\zeta_2 - \sigma_3^{\frac{1}{2}}\zeta_3$$

for any choice of the square roots, and defining β_2, β_3 similarly by cyclic permutation of the suffixes, we have

$$\beta_1 + \beta_2 + \beta_3 = 0. \tag{5}$$

Now β_1 is a non-zero element of the field generated by $\sigma_2^{\frac{1}{2}}$ and $\sigma_3^{\frac{1}{2}}$ over K; further, on multiplying by $\delta = \eta_1\eta_2\eta_3$, one obtains an algebraic integer with field norm having absolute value at most c_6. It follows easily, as above, that $\delta\beta_1 = \beta_1'\epsilon_1^3$ for some unit ϵ_1 in the field and some associate β_1' with $\|\beta_1'\| < c_7$; and, after permutation of suffixes, the same holds for β_2, β_3. Thus (5) gives

$$\beta_1'\epsilon_1^3 + \beta_2'\epsilon_2^3 + \beta_3'\epsilon_3^3 = 0,$$

and, on multiplying by $\beta_2'^2/\epsilon_3^3$, this becomes a Thue equation

$$x^3 - \lambda y^3 = \mu,$$

where $$x = \beta_2'\epsilon_2/\epsilon_3, \quad y = \epsilon_1/\epsilon_3.$$

Hence, by Theorem 4.1, $\|x\|$ and $\|y\|$ are at most c_8, and it remains only to show that $\|X\|$ and $\|Y\|$ are likewise bounded.

Fixing the choice of the sign of $\sigma_2^{\frac{1}{2}}$, one can plainly select the sign of $\sigma_3^{\frac{1}{2}}$ in β_3 so that $|\epsilon_3| < c_9$. Then the bound $|y| < c_8$ established above gives $|\epsilon_1| < c_{10}$, whence, since $|\delta| > c_{11}$, we obtain $|\beta_1| < c_{12}$. But this holds for either choice of the sign of $\sigma_3^{\frac{1}{2}}$ and thus we conclude that both $|\zeta_2|$ and $|\zeta_3|$ are at most c_{13}. It is now apparent from (4) that $|X| < c_{14}$; on commencing with the equations conjugate to (3) we derive the same bound for each conjugate of X, and the theorem follows.

4. Curves of genus 1

Let $f(x,y)$ be an absolutely irreducible polynomial with integer coefficients such that the curve $f(x,y) = 0$ has genus 1. We prove:

Theorem 4.3. *The equation $f(x,y) = 0$ has only a finite number of solutions in integers x, y and these can be effectively determined.*

As mentioned in § 1, the first part of the theorem was initially established by Siegel in 1929, but his method of proof was ineffective. The argument we shall give here, which is based on a birational transformation that reduces the equation to the canonical form considered in Theorem 4.2, provides an effective and simpler proof of Siegel's theorem in the case of curves of genus 1; but it does not seem to extend easily to curves of higher genus.

We shall denote by Ω, $\Omega(x)$ and K respectively the field of all algebraic numbers, the field of rational functions in x with coefficients in Ω, and the finite algebraic extension of $\Omega(x)$ formed by adjoining a root of $f(x,y) = 0$. By the Riemann–Roch theorem, there exist rational functions X_1, X_2 on the curve with orders -2, -3 respectively at some fixed infinite valuation, say Q, of K, and with non-negative orders at all other valuations of K; moreover, one can effectively determine the algebraic coefficients in their Puiseux expansions. We now observe, following Chevalley, that the seven functions 1, X_1, X_2, X_1^2, X_2^2, X_1^3, X_1X_2 have poles of order at most 6 at Q and so, by the Riemann–Roch theorem again, they are linearly dependent over Ω.

Let $p_1, \ldots, p_7$ be the respective coefficients in the linear equation relating them; clearly we have $p_5 \neq 0$, for the six functions excluding X_2^2 have distinct orders at Q. On writing

$$X = X_1, \quad Y = 2p_5 X_2 + p_7 X_1 + p_3,$$

we obtain

$$Y^2 = aX^3 + bX^2 + cX + d,$$

where a, b, c, d are polynomials in $p_1, \ldots, p_7$ with integer coefficients. The cubic on the right has distinct zeros, for if the equation reduced to

$$\{Y/(X-\alpha)\}^2 = a(X-\beta),$$

then $Y/(X-\alpha)$ could possess a pole only at Q; but, since X_1, X_2 have orders -2, -3 respectively at Q and $p_5 \neq 0$, the function has in fact a pole of order 1 at Q, contrary to the Riemann–Roch theorem.

We observe now that, since X_1, X_2 are rational functions of x, y with coefficients in a fixed field, the functions X, Y become algebraic numbers in a fixed field when x, y are rational integers. Moreover, there exists a non-zero rational integer q, independent of x and y, such that qX and qY are algebraic integers; for the function $X = X_1$ has a pole only at the infinite valuation Q and thus the equation satisfied by X over $\Omega(x)$ has the form

$$X^m + P_1(x)\,X^{m-1} + \ldots + P_m(x) = 0,$$

where m is the degree of f in y and $P_1, \ldots, P_m$ are polynomials in x with algebraic coefficients and degree at most 2. We conclude from Theorem 4.2 that X, Y can take only finitely many values when x, y are rational integers. On noting again that X has a pole at Q, it follows at once that there are only finitely many x, and, in view of the initial equation $f(x,y) = 0$, so also finitely many y. Further, it is readily confirmed that all the arguments employed above are, in principle, effective, and this proves Theorem 4.3.

The method of proof can easily be extended to treat curves of genus 0 when there exist at least three infinite valuations, and this together with the above result enables one to resolve effectively the general cubic equation $f(x,y) = 0$; the latter can, however, be reduced more directly to the form considered in Theorem 4.2.

5. Quantitative bounds

As remarked earlier, the arguments employed here enable one to furnish explicit upper bounds for the size of all the solutions of the above equations. To calculate these bounds one needs first a quantitative version of Theorem 3.1, and, in this connexion, the most useful result[†] so far established reads:

If $\alpha_1, \ldots, \alpha_n$ *are* $n \geqslant 2$ *non-zero algebraic numbers with degrees and heights at most* d $(\geqslant 4)$ *and* A $(\geqslant 4)$ *respectively, and if rational integers* $b_1, \ldots, b_n$ *exist with absolute values at most* B *such that*

$$0 < |b_1 \log \alpha_1 + \ldots + b_n \log \alpha_n| < e^{-\delta B},$$

where $0 < \delta \leqslant 1$, *and the logarithms have their principal values, then*

$$B < (4^{n^2} \delta^{-1} d^{2n} \log A)^{(2n+1)^2}.$$

By applying this together with certain estimates for units in algebraic number fields, it has been shown that all solutions X, Y of the Thue equation referred to in Theorem 4.1 satisfy

$$\max(\|X\|, \|Y\|) < \exp\{(dH)^{(10d)^5}\},$$

where H denotes the maximum of the heights of $\alpha_1, \ldots, \alpha_n$, μ and some algebraic integer generating K.[‡] This leads to the bound

$$\exp\exp\exp(d^{10d^3} H^{d^3})$$

for the sizes of all solutions X, Y of the hyperelliptic equation discussed in Theorem 4.2. Further, employing the latter estimate and an effective construction for rational functions,[§] it has been proved that all integer points x, y on the curve $f(x, y) = 0$ of Theorem 4.3 satisfy

$$\max(|x|, |y|) < \exp\exp\exp\{(2H)^{10^{n^{10}}}\},$$

where H denotes the maximum of the absolute values of the coefficients of f and n denotes its degree.[‖]

In special cases one has stronger bounds. For instance, for the elliptic equation mentioned after the enunciation of Theorem 4.2, the estimate

$$\max(|x|, |y|) < \exp\{(10^6 H)^{10^6}\}$$

has been established, where a, b, c, d are assumed to have absolute values at most H; and for the Mordell equation $y^2 = x^3 + k$, it has been shown, by way of an expression for C in terms of Ω similar to that recorded after Theorem 3.1, that the bound $\exp(c|k|^{1+\epsilon})$ is valid for any $\epsilon > 0$, where c depends only on ϵ.[¶] Furthermore, techniques have been devised which, for a wide range of numerical examples, render the problem of determining the complete list of solutions in question accessible to machine computation; thus, for example, it has been proved that the only integer solutions of the pair of equations

† *Mathematika*, **15** (1968), 204–16.
‡ *Phil. Trans. Roy. Soc. London*, A **263** (1968), 173–91; *P.C.P.S.* **65** (1969), 439–44.
§ *P.C.P.S.* **68** (1970), 105–23 (J. Coates). ‖ *P.C.P.S.* **67** (1970), 595–602.
¶ *Acta Arith.* **24** (1973), 251–9 (H. Stark).

$3x^2-2=y^2$ and $8x^2-7=z^2$ are given by $x=1$ and $x=11$, and that the equation $y^2=x^3-28$ has only the solutions given by $x=4, 8, 37$ (the corresponding values of y being ± 6, ± 22, ± 225 respectively).†

Much interest attaches to the size of the solutions of the original Thue equation $F(x,y)=m$ (see §1) relative to m. As a consequence of the third inequality for $|\Lambda|$ recorded after the enunciation of Theorem 3.1, the arguments leading to Theorem 4.1 show that, if $m \geqslant 2$, then $|x|$ and $|y|$ cannot exceed m^C for some computable C depending only on F.‡ This yields at once an improvement on Liouville's theorem; indeed, with the notation of Theorem 1.1, we have

$$|\alpha - p/q| > c/q^\kappa$$

for all rationals p/q $(q>0)$, where c, κ are positive numbers, effectively computable in terms of α, with $\kappa < n$. The result, in slightly weaker form, was first established§ in 1967, particular cases, however, having been obtained a few years earlier by means of special properties of Gauss' hypergeometric function.|| For instance it had been proved¶ that when α is the cube-root of 2 and 17 then the above inequality holds with $c=10^{-6}$, $\kappa=2{\cdot}955$ and $c=10^{-9}$, $\kappa=2{\cdot}4$ respectively, values in fact that are almost certainly sharper than those given by the more general techniques. But, leaving aside the effective nature of c, much more about rational approximations to algebraic numbers is known from the field of research begun by Thue, and this will be the theme of Chapter 7.

Various other equations can be treated by the methods described here. They can be used, for instance, to give bounds for all solutions in integers x, y of the equation $y^m=f(x)$, where $m>2$ and f denotes any polynomial with integer coefficients possessing at least two distinct zeros; in particular, they enable one to solve effectively the Catalan equation $x^m-y^n=1$ for any given m, n.†† Moreover, they can be generalized by means of analysis in the p-adic domain to furnish all rational solutions of the equations $F(x,y)=m$ and $y^2=x^3+k$ whose denominators are comprised solely of powers of fixed sets of primes; thus, more especially, they yield an effective determination of all elliptic curves with a given conductor.‡‡

† *Quart. J. Math. Oxford Ser.* (2) **20** (1969), 129–37; *J. Number Th.* **4** (1972), 107–17.
‡ *I.A.N.* **35** (1971), 973–90.
§ *Phil. Trans. Roy. Soc. London*, A **263** (1968), 173–91.
|| *Proc. London Math. Soc.* **4** (1964), 385–98.
¶ *Quart. J. Math. Oxford Ser.* (2) **15** (1964), 375–83.
†† *P.C.P.S.* **65** (1969), 439–44. In fact R. Tijdeman has recently shown that they enable one to give an effective bound for *all* solutions x, y, m, n of the Catalan equation.
‡‡ *Acta Arith.* **15** (1969), 279–305; **16** (1970), 399–412, 425–35 (J. Coates).

5

CLASS NUMBERS OF IMAGINARY QUADRATIC FIELDS

1. Introduction

The foundations of the theory of binary quadratic forms, the forerunner of our modern theory of quadratic fields, were laid by Gauss in his famous *Disquisitiones Arithmeticae*. Gauss showed, amongst other things, how one could divide the set of all binary quadratic forms into classes such that two forms belong to the same class if and only if there exists an integral unimodular substitution relating them, and he showed also how one could combine the classes into genera so that two forms are in the same genus if and only if they are rationally equivalent. He also raised a number of notorious problems; in particular, in Article 303, he conjectured that there are only finitely many negative discriminants associated with any given class number, and moreover that the tables of discriminants which he had drawn up in the cases of relatively small class numbers were in fact complete. The first part of the conjecture was proved, after earlier work of Hecke, Mordell and Deuring, by Heilbronn[†] in 1934, and the techniques were later much developed by Siegel and Brauer to give a general asymptotic class number formula; but the arguments are non-effective and cannot lead to a verification of the class number tables as sought by Gauss. In 1966, two distinct algorithms were discovered for determining all the imaginary quadratic fields with class number 1, which amounts to a confirmation of the simplest case of the second part of the conjecture.

Theorem 5.1. *The only imaginary quadratic fields* $Q(\sqrt{(-d)})$ *with class number* 1, *where* d *is a square-free positive integer, are given by* $d = 1, 2, 3, 7, 11, 19, 43, 67, 163$.

One of the original methods of proof, and that which we shall adopt here, is based on the work of Chapters 2 and 3 together with an idea of Gelfond and Linnik;[‡] the other is due to Stark[§] and is motivated by an earlier paper of Heegner[||] which related the problem to the study of

† *Quart. J. Math. Oxford Ser.* **5** (1934), 150–60.
‡ *D.A.N.* **61** (1948), 773–6.
§ *Michigan Math. J.* **14** (1967), 1–27. || *M.Z.* **56** (1952), 227–53.

elliptic modular functions and the solution of certain Diophantine equations. The former method has recently been extended to resolve the analogous problem for class number 2, and we shall describe the solution in § 5. Neither method, however, would seem to generalize readily to higher class numbers.

Nevertheless, transcendental number theory has led to new results in several associated subjects. For instance, it has been used by Anferteva and Chudakov† to make effective certain theorems of Linnik on the average of the minimum of the norm function over ideals in a given class, and it has been employed by Schinzel and the author in studies relating to the '*numeri idonei*' of Euler.‡ Furthermore, it has been applied to resolve in the negative a well-known problem of Chowla as to whether there exists a rational-valued function $f(n)$, periodic with prime period p, such that $\Sigma f(n)/n = 0$.§ In fact it has provided a description of all such functions f that take algebraic values and are periodic with any modulus q; thus, in particular, it has revealed that the numbers $L(1,\chi)$ taken over all non-principal characters $\chi \pmod q$ are linearly independent over the rationals, provided only that $(q, \phi(q)) = 1$, and this plainly generalizes Dirichlet's famous result on the non-vanishing of $L(1,\chi)$. It would be of interest to know whether the theorem remains valid when

$$(q, \phi(q)) > 1.$$

Some further results will be mentioned in § 5.

2. *L*-functions

We record here some preliminary observations on products of Dirichlet's *L*-functions.

Let $-d < 0$ and $k > 0$ denote the discriminants of the quadratic fields $Q(\sqrt{(-d)})$ and $Q(\sqrt{k})$ respectively, and suppose that $(k,d) = 1$. Let

$$\chi(n) = \left(\frac{k}{n}\right), \quad \chi'(n) = \left(\frac{-d}{n}\right)$$

be the usual Kronecker symbols. Then, for any $s > 1$, we have

$$L(s,\chi)\,L(s,\chi\chi') = \tfrac{1}{2}\sum_{f}\sum_{x,y}\chi(f)f^{-s}, \tag{1}$$

where x, y run through all integers, not both 0, and

$$f = f(x,y) = ax^2 + bxy + cy^2$$

† *Mat. Sb.* **82** (1970), 55–66; = **11** (1970), 47–58.
‡ *Acta Arith.* **18** (1971), 137–44.
§ *J. Number Th.* **5** (1973), 224–36.

runs through a complete set of inequivalent quadratic forms with discriminant $-d$. To verify this assertion, we observe that the left-hand side of (1) is given by

$$\sum_{m=1}^{\infty}\sum_{n=1}^{\infty}\left(\frac{k}{m}\right)\left(\frac{-kd}{n}\right)(mn)^{-s} = \sum_{l=1}^{\infty}\left(\frac{k}{l}\right)l^{-s}\sum_{n|l}\left(\frac{-d}{n}\right),$$

and the last sum is one half the number of representations of l by the forms f.†

Now the right-hand side of (1) can be written

$$\sum_{f}\sum_{x=1}^{\infty}\chi(ax^2)(ax^2)^{-s} + \sum_{f}\sum_{y=1}^{\infty}\sum_{x=-\infty}^{\infty}\chi(f)f^{-s}.$$

The first term here is

$$\zeta(2s)\prod_{p|k}(1-p^{-2s})\sum_{f}\chi(a)a^{-s},$$

and the second term can be expanded as a Fourier series

$$\sum_{f}\sum_{r=-\infty}^{\infty}A_r(s)\,e^{\pi irb/(ka)},$$

where
$$A_r(s) = k^{-1}\int_0^k\sum_{y=1}^{\infty}\sum_{x=-\infty}^{\infty}\chi(f)\,g^{-s}e^{-2\pi irv/k}\,dv,$$

and
$$g = g(v) = a(x+vy)^2+(d/4a)\,y^2,$$

so that $f = g(b/2a)$. On substituting u for v by the equation

$$x+vy = uy(\sqrt{d}/2a),$$

writing $x = m+kyn$, where $0 \leqslant m < ky$, and interchanging the order of integration and summation, as one may by dominated convergence, one obtains

$$A_r(s) = k^{-1}a^{-s}(\sqrt{d}/2a)^{1-2s}I_r(s)\sum_{y=1}^{\infty}\sigma(y)\,y^{-2s},$$

where
$$I_r(s) = \int_{-\infty}^{\infty}\frac{e^{-\pi iur\sqrt{d}/(ka)}}{(u^2+1)^s}\,du$$

and
$$\sigma(y) = \sum_{m=0}^{ky-1}\chi(f(m,y))\,e^{2\pi irm/(ky)};$$

the integral in fact arises from summation over n of the partial integrals from c_n to c_{n+1}, where

$$c_n = 2a(m+kyn)/(y\sqrt{d}).$$

† See Landau's *Vorlesungen über Zahlentheorie* (Leipzig, 1927), Satz 204.

On putting $m = j+kl$, where $1 \leqslant j \leqslant k$, one sees that

$$\sigma(y) = y\sum_{j=1}^{k}\chi(f(j,y))\,e^{2\pi irj/(ky)}$$

if y divides r, and $\sigma(y) = 0$ otherwise, and this completes the preliminary observations.

3. Limit formula

All solutions to date of the class number 1 problem depend on an analogue for products of L-functions of the classical Kronecker limit formula. On writing, with the notation of the previous section,

$$A_0 = \lim_{s\to 1} A_0(s), \quad A_r = A_r(1) \quad (r \neq 0),$$

and taking limits as $s \to 1$, we obtain

$$L(1,\chi)\,L(1,\chi\chi') = \frac{\pi^2}{6}\prod_{p|k}\left(1-\frac{1}{p^2}\right)\sum_f \frac{\chi(a)}{a} + \sum_f \sum_{r=-\infty}^{\infty} A_r\, e^{\pi irb/(ka)}. \quad (2)$$

Our purpose here is to prove that

$$|A_r| \leqslant \frac{2\pi}{\sqrt{d}}|r|\, e^{-\pi|r|\sqrt{d}/(ka)}$$

for $r \neq 0$, and

$$A_0 = \frac{-2\pi}{k\sqrt{d}}\chi(a)\log p$$

if k is the power of a prime p, $A_0 = 0$ otherwise.

To begin with, we observe that, for $r \neq 0$,

$$A_r = (\tfrac{1}{2}k\sqrt{d})^{-1} I_r(1)\sum_y \sum_{j=1}^{k} y^{-1}\chi(f(j,y))\,e^{2\pi irj/(ky)},$$

where y runs through all positive divisors of r. It is easily confirmed that

$$I_r(1) = \pi e^{-\pi|r|\sqrt{d}/(ka)},$$

and clearly the sum over y in A_r has absolute value at most $k|r|$. The first assertion follows at once. To establish the second assertion, we note that

$$A_0(s) = k^{-1}a^{s-1}(\tfrac{1}{2}\sqrt{d})^{1-2s} I_0(s)\sum_{y=1}^{\infty} y^{1-2s}\sum_{j=1}^{k}\chi(f(j,y)),$$

and

$$I_0(s) = \sqrt{\pi}\,\Gamma(s-\tfrac{1}{2})/\Gamma(s).$$

Further, by well-known estimates for the Gaussian sums, we obtain, for any positive integer y and any odd k,

$$\sum_{j=1}^{k} \chi(f(j,y)) = \chi(a) \sum_{j=1}^{k} \chi(j^2)\, e^{2\pi i j y/k};$$

we shall be concerned in the sequel only with odd values of k, but the equation in fact holds also for even k, as has been shown by Stark.† The sum over j on the right can be expressed alternatively as a sum of terms $d\mu(k/d)$ over all common divisors d of k and y,‡ and hence we see that the sum over y in the above expression for $A_0(s)$ is given by

$$\chi(a)\,\zeta(2s-1)\,k^{2-2s} \prod_{p|k} (1-p^{2s-2}).$$

The required result is now readily verified.

4. Class number 1

Suppose that $Q(\sqrt{(-d)})$ has class number 1. Then, by the theory of genera, d is a prime congruent to 3 (mod 4), and there is just one form f which can be taken as

$$x^2 + xy + \tfrac{1}{4}(1+d)\,y^2.$$

We select $k = 21$ and we note that $Q(\sqrt{k})$ has class number 1 and fundamental unit $\epsilon = \frac{1}{2}(5+\sqrt{21})$. Further we note that $(k,d) = 1$ for $d > k$, and that $A_0 = 0$. Hence the double sum on the right of (2) has absolute value at most

$$(4\pi/\sqrt{d}) \sum_{r=1}^{\infty} r\eta^r,$$

where $\eta = e^{-\pi\sqrt{d}/k}$. The sum over r is precisely $\eta/(1-\eta)^2$, and $\eta < \frac{1}{2}$ if $\sqrt{d} > k$; thus the above expression is at most $16\pi\eta/\sqrt{d}$.

Now classical results of Dirichlet give

$$L(1,\chi) = 2\log\epsilon/\sqrt{k}, \quad L(1,\chi\chi') = h\pi/\sqrt{(kd)},$$

where h denotes the class number of $Q(\sqrt{(-kd)})$, and, on substituting into (2), we readily derive the inequality

$$|h\log\epsilon - \tfrac{32}{21}\pi\sqrt{d}| < e^{-\pi\sqrt{d}/100},$$

assuming that $d > 10^{20}$, say. But $\pi = -2i\log i$ and so we have on the left a linear form Λ in two logarithms of the kind considered

† *Acta Arith.* **14** (1968), 35–50.

‡ See Hardy and Wright's, *An introduction to the theory of numbers* (Oxford, 1960), Theorem 271.

in Theorem 3.1; since clearly $h < 4\sqrt{d}$ and $\log \epsilon$, $\log i$ are linearly independent, we conclude that the inequality is untenable if d is larger than some effectively computable number. To calculate the latter, it is convenient to take a second inequality arising from (2) with $k = 33$, namely

$$|h' \log \epsilon' - \tfrac{80}{33}\pi \sqrt{d}| < e^{-\pi\sqrt{d}/100},$$

where h', ϵ' are defined like h, ϵ above with the new value of k. By subtraction we obtain

$$|b \log \epsilon + b' \log \epsilon'| < e^{-\delta B},$$

where $\delta^{-1} = 14 \times 10^3$, $B = 140\sqrt{d}$, $b = 35h$, $b' = -22h'$,

and clearly b, b' have absolute values at most B. Since, furthermore, ϵ, ϵ' are multiplicatively independent, one can apply the result quoted in § 5 of Chapter 4, with $n = 2$, $d = 4$, $A = 46$, to obtain $B < 10^{250}$. This gives $d < 10^{500}$, and a determination of the solutions of the above inequality below this figure is quite feasible. But the computation is in fact not needed here, for it was proved by Heilbronn and Linfoot† in 1934 that, apart from the nine discriminants listed in Theorem 5.1, there could be at most one more, and calculations‡ had shown that the tenth d, if it existed, would exceed $\exp(10^7)$.

The above argument is similar to that described by Gelfond and Linnik in 1949, but they had access to the formulae of § 3 only for prime values of k, and in this case A_0 is not 0; thus they were led to an inequality involving three logarithms of algebraic numbers which could not be dealt with effectively at that time. It is a remarkable coincidence that both the formulae for composite k and the desired effective inequality involving three logarithms became available simultaneously in 1966.

5. Class number 2

We now indicate briefly how the above arguments can be extended to treat the analogous problem for class number 2.§

If $Q(\sqrt{(-d)})$ has class number 2 and $d > 15$ then d is congruent to 3 or 4 (mod 8); for if $d \equiv 7 \pmod 8$ there are three inequivalent quadratic forms with discriminant $-d$, namely

$$x^2 + xy + \tfrac{1}{4}(1+d)y^2, \quad 2x^2 \pm xy + \tfrac{1}{8}(1+d)y^2.$$

† *Quart. J. Math. Oxford Ser.* 5 (1934), 293–301.
‡ *Trans. Amer. Math. Soc.* 122 (1966), 112–19 (H. M. Stark).
§ For the original solutions see *Ann. Math.* 94 (1971), 139–52 (A. Baker); 153–73 (H. M. Stark).

When $d \equiv 4 \pmod 8$, two inequivalent quadratic forms with discriminant $-d$ are given by $x^2+\frac{1}{4}dy^2$, and either

$$2x^2+2xy+\tfrac{1}{8}(4+d)\,y^2 \quad \text{or} \quad 2x^2+\tfrac{1}{8}dy^2,$$

according as $\frac{1}{4}d \equiv 1$ or $2 \pmod 4$, and the method of proof of Theorem 5.1 is applicable with only simple modifications.† There remains the case $d \equiv 3 \pmod 8$. The theory of genera shows that then $d = pq$, where p, q are primes congruent to 1 and 3 (mod 4) respectively. On signifying by $\chi'(n)$ one of the generic characters associated with forms of discriminant $-d$ and writing

$$\chi_{pq}(n) = \left(\frac{-pq}{n}\right), \quad \chi_p(n) = \left(\frac{p}{n}\right), \quad \chi_q(n) = \left(\frac{-q}{n}\right), \quad \chi(n) = \left(\frac{k}{n}\right),$$

where $k \equiv 1 \pmod 4$ and $(k, pq) = 1$, we deduce from classical results of Dirichlet and Kronecker that

$$L(1,\chi)\,L(1,\chi\chi_{pq})+L(1,\chi\chi_p)\,L(1,\chi\chi_q)$$
$$=\tfrac{1}{2}\sum_F \sum_{x,y} (\chi(F)+\chi\chi'(F))\,(F(x,y))^{-1},$$

where F runs through a pair f, f' of inequivalent quadratic forms with discriminant $-d$ and x, y take all integer values, not both 0. We can assume that f is the principal form, whence $\chi'(f) = 1$, $\chi'(f') = -1$ for all x, y. On appealing to Dirichlet's formulae we thus obtain

$$(k/2\pi)\sqrt{(pq)}\sum_{x,y}\chi(f)/f = h(k)\,h(-kpq)\log\epsilon + h(kp)\,h(-kq)\log\eta,$$

where $h(l)$ denotes the class number of $Q(\sqrt{l})$ and ϵ, η denote the fundamental units in $Q(\sqrt{k})$, $Q(\sqrt{(kp)})$ respectively. Finally taking $k = 21$ and employing arguments similar to those applied in the proof of Theorem 5.1, we reach the inequality

$$|h(-21d)\log\epsilon + h(21p)\,h(-21q)\log\eta - \tfrac{64}{21}\pi\sqrt{d}| < e^{(-1/10)\sqrt{d}}.$$

This has the form

$$|\beta\log\alpha+\beta'\log\alpha'+\beta''\log\alpha''| < e^{-\delta B},$$

where the β's denote algebraic numbers with degrees at most 2, and $\alpha = \eta, \alpha' = \epsilon$, $\alpha'' = -1$, $B = \sqrt{d}$, $\delta = \frac{1}{10}$. Clearly the heights of the β's are bounded above by an absolute power of B and the height A of α is bounded above by $p^{c\sqrt{p}}$ for some absolute constant c. If $q \leqslant d^{\frac{1}{2}}$ then we can take f' as

$$qx^2+qxy+\tfrac{1}{4}(p+q)\,y^2,$$

† See *Bull. Lond Math. Soc.* 1 (1969), 98–102.

and again the method of proof of Theorem 5.1 is applicable. Thus we can assume that $q > d^{\frac{1}{4}}$ whence $p < d^{\frac{3}{4}}$. We now appeal to the first inequality for $|\Lambda|$ recorded after the enunciation of Theorem 3.1 and, on noting that the maximum A' of the heights of α', α'' is absolutely bounded, we conclude that $B < C(\log A)^{1+\zeta}$ for any $\zeta > 0$, where $C = C(\zeta)$ is effectively computable. Hence we have

$$\sqrt{d} < C(c\sqrt{p}\log p)^{1+\zeta}$$

and, recalling that $p < d^{\frac{3}{4}}$, this plainly gives an effective upper estimate for d when $\zeta < \frac{1}{3}$. In practice† the bound for d turns out to be a little over 10^{1000}, and computational work on the zeros of the ζ-function has yielded all d in question below this figure; thus it has been checked that the largest d for which $Q(\sqrt{(-d)})$ has class number 2 is 427.

Progress in this and other fields of application of the theory of linear forms in the logarithms of algebraic numbers is continuing, and, before leaving the topic, we record five further results that have been obtained with its aid. First it has been utilized by E. E. Whitacker‡ to determine certain imaginary quadratic fields with the Klein four-group as class group. Secondly it has been employed by K. Ramachandra and T. N. Shorey§ in researches on a problem of Erdös in prime-number theory; in particular, they have shown that if k is a natural number and if $n_1, n_2, \ldots$ is the sequence, in ascending order, of all natural numbers which have at least one prime factor exceeding k, then the maximum $f(k)$ of $n_{i+1}-n_i$ $(i = 1, 2, \ldots)$ satisfies $f(k)\log k/k \to 0$ as $k \to \infty$. Thirdly, in a similar context, R. Tijdeman‖ has used an inequality for $|\Lambda|$ of the kind appearing after Theorem 3.1 to resolve in the affirmative a question of Wintner as to whether there exists a sequence of primes such that the sequence $n_1, n_2, \ldots$ of all natural numbers formed from their power products satisfies $n_{i+1}-n_i \to \infty$ as $i \to \infty$. Fourthly, A. Schinzel¶ has applied the second inequality for $|\Lambda|$ recorded after Theorem 3.1 to settle an old problem concerning primitive prime factors of $\alpha^n - \beta^n$. And, finally, we mention that in 1967, A. Brumer†† obtained a natural p-adic analogue of an early version of Theorem 3.1 which, in combination with work of Ax,‡‡ resolved a well-known problem of Leopoldt on the non-vanishing of the p-adic regulator of an Abelian number field.

† *Ann. Math.* **96** (1972), 174–209 (H. M. Stark).
‡ Ph.D. Thesis, University of Maryland, 1972.
§ *Acta Arith.* **24** (1973), 99–111; **25** (1974), 365–73.
‖ *Compositio Math.* **26** (1973), 319–30.
¶ *J.M.* **269** (1974), 27–33.
†† *Mathematika*, **14** (1967), 121–4.
‡‡ *Illinois J. Math.* **9** (1965), 584–9.

6

ELLIPTIC FUNCTIONS

1. Introduction

Siegel[†] proved in 1932 that if $\wp(z)$ is a Weierstrass $\wp$-function such that the invariants g_2, g_3 in the equation

$$(\wp'(z))^2 = 4(\wp(z))^3 - g_2\wp(z) - g_3$$

are algebraic numbers, then one at least of any fundamental pair ω, ω' of periods of $\wp(z)$ is transcendental; thus both ω and ω' are transcendental if $\wp(z)$ admits complex multiplication. Siegel's work was much improved by Schneider[‡] in 1937; Schneider showed that if g_2, g_3 are algebraic then any period of $\wp(z)$ is transcendental, and moreover the quotient ω/ω' is transcendental except in the case of complex multiplication. From the latter result it follows at once that the elliptic modular function $j(z)$ is transcendental for any algebraic z other than an imaginary quadratic irrational. Schneider's work led, in fact, to a wide variety of theorems on the transcendence of values of the Weierstrass functions, and, in 1941, he further obtained far-reaching generalizations concerning Abelian functions and integrals.[§]

Most of Schneider's results in this context can be derived as particular cases of a general theorem on meromorphic functions which he proved in 1949.[‖] The theorem has recently been re-formulated by Lang.[¶]

Theorem 6.1. *Let K be an algebraic number field and let $f_1(z), \ldots, f_n(z)$ be meromorphic functions of finite order. Suppose that the ring $K[f_1, \ldots, f_n]$ is mapped into itself by differentiation and has transcendence degree at least 2 over K. Then there are only finitely many numbers z at which $f_1, \ldots, f_n$ simultaneously assume values in K.*

A meromorphic function $f(z)$ is said to have finite order if there exists $\rho > 0$ and a representation of f as a quotient g/h of entire functions such that, for any $R \geqslant 2$, and for all z with $|z| \leqslant R$, one has

$$\max(|g(z)|, |h(z)|) < \exp(R^\rho). \tag{1}$$

† *J.M.* **167** (1932), 62–9. ‡ *M.A.* **113** (1937), 1–13.
§ *J.M.* **183** (1941), 110–28. ‖ *M.A.* **121** (1949), 131–40.
¶ See Bibliography (first work).

The ring $K[f_1, \ldots, f_n]$ consists of all polynomials in $f_1, \ldots, f_n$ with coefficients in K, and the transcendence degree is the maximum number of elements in an algebraically independent subset. Theorem 6.1 has been generalized to relate to meromorphic functions of several variables but the assertion has been obtained only for point sets which can be represented essentially as a cartesian product and this limits considerably the range of application.† Functions of several variables have been utilized, however, as in Chapters 2 and 3, in other work on elliptic functions, and this will be the theme of §6.

2. Corollaries

We now record some corollaries to Theorem 6.1; others can be found in the works cited in the Bibliography.

Theorem 6.2. *If g_2, g_3 are algebraic, then for any algebraic $\alpha \neq 0$, $\wp(\alpha)$ is transcendental.*

For the proof one has merely to observe that if $\wp(\alpha)$ were algebraic then, for infinitely many integral values of z, the functions

$$f_1(z) = \wp(\alpha z), \quad f_2(z) = \wp'(\alpha z), \quad f_3(z) = z$$

would simultaneously assume values in the algebraic number field generated by g_2, g_3, α, $\wp(\alpha)$ and $\wp'(\alpha)$ over the rationals, contrary to Theorem 6.1.

Theorem 6.3. *For any algebraic α with positive imaginary part, other than a quadratic irrational, $j(\alpha)$ is transcendental.*

For suppose that $j(\alpha)$ is algebraic. Then there is a $\wp$-function with algebraic invariants g_2, g_3 and fundamental periods ω_1, ω_2 such that $\alpha = \omega_2/\omega_1$; indeed if $\bar{\wp}(z)$ is the $\wp$-function with periods 1, α and if $\bar{g}_2$, $\bar{g}_3$ are the invariants of $\bar{\wp}$ then the required $\wp$-function has periods $\bar{g}_3^{\frac{1}{6}}$, $\alpha\bar{g}_3^{\frac{1}{6}}$ if $\bar{g}_3 \neq 0$ and $\bar{g}_2^{\frac{1}{4}}$, $\alpha\bar{g}_2^{\frac{1}{4}}$ if $\bar{g}_2 \neq 0$. Now the functions $f_1 = \wp(z)$, $f_2 = \wp(\alpha z)$, $f_3 = \wp'(z)$, $f_4 = \wp'(\alpha z)$ simultaneously assume values in an algebraic number field, say K, when $z = (r+\frac{1}{2})\,\omega_1$ $(r = 1, 2, \ldots)$ and so, by Theorem 6.1, $K[f_1, f_2, f_3, f_4]$ has transcendence degree at most 1. This implies that f_1, f_2 are algebraically dependent, whence $l\omega_2$ is a period of $\wp(\alpha z)$ for some positive integer l. Thus $l\alpha\omega_2 = m\omega_1 + n\omega_2$ for some integers m, n and so α is a quadratic irrational. It will be recalled that if 1, α is a basis for an imaginary quadratic field K, then $j(\alpha)$ is in fact a real algebraic integer with degree given by the class number of

† For some work aimed towards overcoming this difficulty see papers by Bombieri (*Invent. Math.* **10** (1970), 267–87) and Bombieri and Lang (*ibid.* **11** (1970), 1–14). It is shown that it suffices if the points in question do not lie on an algebraic hypersurface.

K, and hence the hypothesis of Theorem 6.3 is certainly necessary.

Theorem 6.4. *Any vector period of an Abelian function arising from an algebraic curve*† *by the inversion of Abelian integrals is transcendental.*

The result follows from Theorem 6.1 with $f_1(z), \ldots, f_{n-1}(z)$ given by the Abelian function, say $A(z_1, \ldots, z_p)$, and its p partial derivatives with respect to $z_1, \ldots, z_p$, evaluated at $z_1 = \omega_1 z, \ldots, z_p = \omega_p z$, where $(\omega_1, \ldots, \omega_p)$ denotes the given period, together with $f_n(z) = z$. It should perhaps be emphasized that the theorem establishes only the transcendence of one at least of the elements of the period vector, and it remains an open problem to prove the transcendence of each such element. An analogue of Theorem 6.4 was obtained by Schneider‡ in 1941; he showed, by many-variable techniques, that one at least of the first elements of the $2p$ fundamental periods is transcendental and this implies that the β-function

$$\beta(a, b) = \int_0^1 x^{a-1}(1-x)^{b-1}\,dx = \frac{\Gamma(a)\,\Gamma(b)}{\Gamma(a+b)}$$

is transcendental for all rational, non-integral a, b. For if $a+b$ is not an integer then the first elements of the periods of the Abelian function arising from the integration of $x^{a-1}(1-x)^{b-1}$ are given by products of $\beta(a, b)$ with numbers in the field generated by $e^{2\pi i a}$ and $e^{2\pi i b}$ over the rationals; and the case when $a+b$ is an integer reduces to the transcendence of π. This result on $\beta(a, b)$ represents all that is known concerning the transcendence of the values of the Γ-function.

Finally, let ω be a primitive period of a $\wp$-function with algebraic invariants g_2, g_3 and let $\eta = 2\zeta(\frac{1}{2}\omega)$ be the associated quasi-period of the Weierstrass ζ-function satisfying $\zeta'(z) = -\wp(z)$. We have

Theorem 6.5. *Any linear combination of* ω, η *with algebraic coefficients, not both* 0, *is transcendental.*

For the proof we observe simply that if $\alpha\omega + \beta\eta$ were algebraic, where α, β are algebraic numbers, not both 0, then the functions

$$f_1 = \wp(z), \quad f_2 = \wp'(z), \quad f_3 = \alpha z + \beta\zeta(z)$$

would simultaneously assume values in an algebraic number field when $z = (r + \frac{1}{2})\,\omega$ $(r = 1, 2, \ldots)$, contrary to Theorem 6.1. On recalling that ω and η can be represented as elliptic integrals of the first and second kinds respectively, one deduces easily from Theorem 6.5 that the circumference of any ellipse with algebraic axes-lengths is transcendental. Further work in this context will be discussed in § 6.

† The curve is defined over the algebraic numbers.
‡ *J.M.* 183 (1941), 110–28.

3. Linear equations

We establish here a result on linear equations with algebraic coefficients which generalizes Lemma 1 of Chapter 2. K will signify an algebraic number field and c_1, c_2, c_3 will denote positive numbers that depend on K only. Further, as in Chapter 4, $\|\theta\|$ will signify the size of θ, that is, the maximum of the absolute values of the conjugates of θ.

Lemma 1. *Let M, N be integers with $N > M > 0$ and let*

$$u_{ij} \ (1 \leqslant i \leqslant M, 1 \leqslant j \leqslant N)$$

be algebraic integers in K with sizes at most U ($\geqslant 1$). Then there exist algebraic integers $x_1, \ldots, x_N$ in K, not all 0, satisfying

$$\sum_{j=1}^{N} u_{ij} x_j = 0 \quad (1 \leqslant i \leqslant M)$$

and
$$\|x_j\| \leqslant c_1 (c_1 N U)^{M/(N-M)} \quad (1 \leqslant j \leqslant N).$$

For the proof we denote by $\omega_1, \ldots, \omega_n$ an integral basis for K and we observe that

$$u_{ij} \omega_k = \sum_{h=1}^{n} u_{hijk} \omega_h$$

for some rational integers u_{hijk}. The equations serve to express the latter as linear combinations of the u_{ij} and their conjugates, with coefficients that depend only on K, and hence we have $|u_{hijk}| < c_2 U$. It follows from Lemma 1 of Chapter 2 that there exist rational integers x_{jk}, not all 0, with absolute values at most $(c_3 N U)^{M/(N-M)}$, satisfying

$$\sum_{j=1}^{N} \sum_{k=1}^{n} u_{hijk} x_{jk} = 0 \quad (1 \leqslant h \leqslant n, 1 \leqslant i \leqslant M),$$

and it is now clear that the numbers

$$x_j = \sum_{k=1}^{n} x_{jk} \omega_k \quad (1 \leqslant j \leqslant N)$$

have the required properties.

4. The auxiliary function

We assume now that the hypotheses of Theorem 6.1 are satisfied and we write $f_i = g_i/h_i$, where g_i, h_i are entire functions for which (1) holds. We suppose further that there exists a sequence of distinct complex numbers $y_1, y_2, \dots$ such that $f_i(y_j)$ is an element of K for all i, j. By $c_4, c_5, \dots$ we shall denote positive numbers which depend only on the quantities so far defined. We signify by m an integer that exceeds a sufficiently large c_4, and by k an integer that is sufficiently large compared with m. We write, for brevity, $L = [k^{3/4}]$, and we use $f^{(j)}$ to denote the jth derivative of f.

Lemma 2. *There are algebraic integers $p(\lambda_1, \lambda_2)$ in K, not all 0, with sizes at most $k^{c_5 k}$, such that the function*

$$\Phi(z) = \sum_{\lambda_1=0}^{L} \sum_{\lambda_2=0}^{L} p(\lambda_1, \lambda_2)\, (f_1(z))^{\lambda_1}\, (f_2(z))^{\lambda_2}$$

satisfies

$$\Phi^{(j)}(y_l) = 0 \quad (0 \leqslant j \leqslant k,\ 1 \leqslant l \leqslant m).$$

Proof. The number $\Phi^{(j)}(y_l)$ is plainly expressible as a linear form in the $p(\lambda_1, \lambda_2)$ with coefficients given by polynomials in $f_1(y_l), \dots, f_n(y_l)$. The polynomials arise from the derivatives of $f_1, \dots, f_n$ which, by hypothesis, are elements of $K[f_1, \dots, f_n]$; thus the coefficients of $p(\lambda_1, \lambda_2)$ belong to K. The latter become algebraic integers when multiplied by some positive integer, and we shall suppose that the sizes of these algebraic integers are at most U. The number of equations to be satisfied is $M = m(k+1)$ and the number of unknowns $p(\lambda_1, \lambda_2)$ is $N = (L+1)^2 > k^{3/2}$. But clearly $N > 2M$ for k sufficiently large and so, by Lemma 1, the equations can be solved non-trivially, and indeed with the sizes of the $p(\lambda_1, \lambda_2)$ at most $c_1^2 NU$. Hence it remains only to prove that one can take $U \leqslant k^{c_6 k}$.

Now it is readily verified by induction on j that, for any polynomial

$$Q(x_1, \dots, x_n) = \sum_{l_1=0}^{d} \dots \sum_{l_n=0}^{d} q(l_1, \dots, l_n)\, x_1^{l_1} \dots x_n^{l_n}$$

with coefficients in K, the function $R(z) = Q(f_1, \dots, f_n)$ satisfies

$$R^{(j)}(z) = \sum_{l_1=0}^{d'} \dots \sum_{l_n=0}^{d'} r(l_1, \dots, l_n) f_1^{l_1} \dots f_n^{l_n},$$

where the $r(l_1, \dots, l_n)$ are again elements of K and $d' \leqslant d + j\delta$, δ denoting the maximum of the degrees of the first derivatives of $f_1, \dots, f_n$, expressed as polynomials in the latter. Further, it is easily confirmed that if the $q(l_1, \dots, l_n)$ become algebraic integers with sizes at most s after multiplying Q by some positive integer, then $R^{(j)}$ can be multiplied by a positive integer so that the $r(l_1, \dots, l_n)$ become algebraic integers with sizes at most $S = (c_7 d)^j j!\, s$. The lemma follows on applying this result with $Q = x_1^{\lambda_1} x_2^{\lambda_2}$ and $j \leqslant k$, whence $s = 1, d \leqslant L \leqslant k$ and $S \leqslant k^{c_9 k}$, and noting that, if k is sufficiently large, then the estimate $k^{c_9 k}$ obtains for each power product $f_1^{l_1} \dots f_n^{l_n}$ evaluated at $z = y_l$, where $l_i \leqslant d' \leqslant c_{10} k$ and $l \leqslant m$.

Lemma 3. *For any $R \geqslant 2$ and for all z with $|z| \leqslant R$, the function $\phi = (h_1 \dots h_n)^L \Phi$ satisfies*

$$|\phi(z)| < \exp\{c_{11}(k \log k + LR^\rho)\}.$$

Further, for any j, l with $j \geqslant k$, $l \leqslant m$ such that $\Phi^{(i)}(y_l) = 0$ for all $i < j$, the number $\phi^{(j)}(y_l)$ either vanishes or has absolute value at least $j^{-c_{12} j}$.

Proof. The first part is an immediate deduction from (1) together with the estimates occurring in Lemma 2. The second part is obtained by an argument similar to that employed in the proof of Lemma 3 of Chapter 2; one observes that $\Phi^{(j)}(y_l)$ is an element of K and that, for $j \geqslant k$, it becomes an algebraic integer with size at most $j^{c_{13} j}$ when multiplied by some positive integer likewise bounded. Further, by hypothesis, $\Phi^{(j)}(y_l)$ differs from $\phi^{(j)}(y_l)$ only by a factor $(h_1 \dots h_n)^L$ evaluated at $z = y_l$, and the required result now follows from the fact that the norm of a non-zero algebraic integer is at least 1.

5. Proof of main theorem

It suffices to prove that Φ vanishes identically; for this implies that f_1 and f_2 are algebraically dependent and so, since the suffixes can be chosen arbitrarily, $K[f_1, \dots, f_n]$ has transcendence degree at most 1, contrary to hypothesis. The contradiction shows that m is bounded by some c_4 as above, whence the sequence $y_1, y_2, \dots$ must terminate.

The proof will proceed by induction on j; we assume that

$$\Phi^{(i)}(y_l) = 0 \quad (0 \leqslant i < j,\ 1 \leqslant l \leqslant m),$$

and we prove that the same then holds for $i = j$. In view of Lemma 2 we can suppose that $j > k$. Let now C be the circle in the complex plane described in the positive sense with centre the origin and radius $R = j^{1/(4\rho)}$. Further, let

$$F(z) = (z - y_1) \dots (z - y_m),$$

and let l be any integer with $1 \leqslant l \leqslant m$. By Cauchy's residue theorem

$$\frac{\phi^{(j)}(y_l)}{(F'(y_l))^j} = \frac{j!}{2\pi i} \int_C \frac{\phi(z)\, dz}{(z - y_l)(F(z))^j}.$$

Clearly for z on C we have

$$|F(z)| > (\tfrac{1}{2}R)^m > j^{m/(8\rho)},$$

and also $|z-y_l| > \frac{1}{2}R$. Further, we have $LR^\rho \leqslant k^{\frac{3}{4}}j^{\frac{1}{4}} \leqslant j$ and so, by Lemma 3, $|\phi(z)| \leqslant j^{c_{14}j}$. Furthermore, it is obvious that $|F'(y_l)| < j$ for k sufficiently large. Hence we obtain

$$|\phi^{(j)}(y_l)| \leqslant j^{c_{15}j-jm/(8\rho)}.$$

But if $m > 8\rho(c_{12}+c_{15})$ then, in view of Lemma 3, the latter estimate implies that $\phi^{(j)}(y_l) = 0$. Assuming, as plainly one may, that $h_1 \ldots h_n$ does not vanish at $z = y_l$, it follows that $\Phi^{(j)}(y_l) = 0$. Thus, by induction, we conclude that Φ and all its derivatives vanish at $y_1, \ldots, y_m$ whence Φ vanishes identically, as required.

6. Periods and quasi-periods

The work of Siegel, cited at the beginning, was based on the interpolation techniques discovered a few years previously by Gelfond,† and the work of Schneider arose out of further developments of these techniques leading, as mentioned in Chapter 2, to a solution of the seventh problem of Hilbert. The recent advances concerning linear forms in the logarithms of algebraic numbers discussed in earlier chapters have similarly given rise to new results on the transcendental theory of elliptic functions, as we shall now describe.

First, generalizing Theorem 6.5, it has been shown that if ω_1, ω_2 are primitive periods of some, possibly distinct $\wp$-functions both with algebraic invariants, and if η_1, η_2 are the associated quasi-periods of the ζ-functions, we have‡

Theorem 6.6. *Any non-vanishing linear combination of $\omega_1, \omega_2, \eta_1, \eta_2$ with algebraic coefficients is transcendental.*

This establishes, in particular, the transcendence of the sum of the circumferences of two ellipses with algebraic axes-lengths. For the proof of Theorem 6.6 we signify by $\wp_1, \wp_2$ the given $\wp$-functions, by ζ_1, ζ_2 the associated ζ-functions and we assume, as we may without loss of generality, that the corresponding invariants $\frac{1}{4}g_2, \frac{1}{4}g_3$ are algebraic integers. We assume also that there exists a linear relation

$$\alpha_1\omega_1+\alpha_2\omega_2+\beta_1\eta_1+\beta_2\eta_2 = \alpha_0,$$

† See e.g. *Tôhoku Math. J.* **30** (1929), 280–5.
‡ *Göttingen Nachrichten* (1969), No. 16, 145–57.

where $\alpha_0 \neq 0$, α_1, α_2, β_1, β_2 are algebraic numbers, and we ultimately derive a contradiction. We signify by k an integer which exceeds a sufficiently large number c depending only on the α's, β's and the invariants, periods and quasi-periods of the Weierstrass functions, and we write, for brevity, $h = [k^{\frac{1}{10}}]$, $L = [k^{\frac{3}{4}}]$. The argument then rests on the construction of an auxiliary function

$$\Phi(z_1, z_2) = \sum_{\lambda_0=0}^{L} \sum_{\lambda_1=0}^{L} \sum_{\lambda_2=0}^{L} p(\lambda_0, \lambda_1, \lambda_2) (f(z_1, z_2))^{\lambda_0} (\wp_1(\omega_1 z_1))^{\lambda_1} (\wp_2(w_2 z_2))^{\lambda_2},$$

where the $p(\lambda_0, \lambda_1, \lambda_2)$ are integers, not all 0, with absolute values at most k^{10k}, and

$$f(z_1, z_2) = \alpha_1 \omega_1 z_1 + \alpha_2 \omega_2 z_2 + \beta_1 \zeta_1(\omega_1 z_1) + \beta_2 \zeta_2(\omega_2 z_2).$$

The function is constructed to satisfy

$$\Phi_{m_1, m_2}(s + \tfrac{1}{2}, s + \tfrac{1}{2}) = 0$$

for all integers s with $1 \leqslant s \leqslant h$ and all non-negative integers m_1, m_2 with $m_1 + m_2 \leqslant k$, where the suffixes denote partial derivatives as in Chapter 2.

The essence of the proof is an extrapolation algorithm analogous to that described in connexion with linear forms in logarithms, but the order of Φ here is greater than in the earlier work and, to compensate, rational extrapolation points with large denominators are utilized; an important rôle in the discussion is therefore played by the division value properties of the elliptic functions. The counterpart of Lemma 4 of Chapter 2 asserts that, for any integer J between 0 and 50 inclusive, we have

$$\Phi_{m_1, m_2}(s + r/q, s + r/q) = 0$$

for all integers q, r, s with q even, $(r, q) = 1$,

$$1 \leqslant q \leqslant 2h^{\frac{1}{8}J}, \quad 1 \leqslant s \leqslant h^{\frac{1}{8}J+1}, \quad 1 \leqslant r < q,$$

and all non-negative integers m_1, m_2 with $m_1 + m_2 \leqslant k/2^J$. The demonstration proceeds by induction and involves an application of the maximum-modulus principle as in the original lemma. It also utilizes the observation that, apart from a factor $\omega_1^{m_1} \omega_2^{m_2}$, the number on the left of the required equation is algebraic with degree at most $c'q^4$, where c' is defined like c above; and precise estimates for the number and its conjugates are furnished by division value theory. One concludes from the lemma that

$$\Phi_{m_1, m_2}(s + \tfrac{1}{4}, s + \tfrac{1}{4}) = 0 \quad (1 \leqslant s \leqslant L + 1,\ 0 \leqslant m_1, m_2 \leqslant L),$$

which is clearly a system of $(L+1)^3$ linear equations in the same number of variables $p(\lambda_0, \lambda_1, \lambda_2)$; on noting that, for any regular function f, the determinant or order n with the ith derivative of $(f(z))^j$ in the ith row and jth column has value

$$2!\dots n!\,(f'(z))^{\frac{1}{2}n(n+1)},$$

one easily verifies that the system of equations is untenable, and this proves Theorem 6.6.

The special case of the theorem when $\wp_1, \wp_2$ are the same $\wp$-function, say $\wp$, is of particular interest. For then ω_1, ω_2 can be taken as a pair of fundamental periods of $\wp$ and we have the Legendre relation

$$\eta_1\omega_2 - \eta_2\omega_1 = 2\pi i.$$

In this case Coates† *and more recently Masser*‡ *have much extended* the arguments and have proved:

Theorem 6.7. *The space spanned by* $1, \omega_1, \omega_2, \eta_1, \eta_2$ *and* $2\pi i$ *over the algebraic numbers has dimension either* 4 *or* 6 *according as* $\wp$ *does or does not admit complex multiplication.*

The theorem clearly exhibits a non-trivial example of five numbers that are algebraically dependent but linearly independent over the algebraic numbers. Moreover it implies that, when $\wp$ admits complex multiplication, the numbers in question satisfy an algebraic linear relation other than that between the periods; this was discovered by Masser. It takes the form

$$a\eta_2 - c\tau\eta_1 = \gamma\omega_2,$$

where γ is algebraic and a, c are the integers occurring in the equation

$$a + b\tau + c\tau^2 = 0$$

satisfied by $\tau = \omega_1/\omega_2$. A necessary and sufficient condition for γ to be 0 is that either g_2 or g_3 be 0, and thus one deduces that η_1/η_2 is transcendental if and only if neither invariant vanishes. The theorem also shows, for instance, that $\pi+\omega$ and $\pi+\eta$ are transcendental for any period ω of $\wp(z)$ and quasi-period η of $\zeta(z)$. The transcendence of π/ω, incidentally, follows from Theorem 6.1 by way of the functions $\wp(\omega z/\pi)$ and e^{2iz}.

The demonstration of Theorem 6.6 extends easily to establish, under the conditions appertaining to Theorem 6.7, the transcendence of any

† *Amer. J. Math.* **93** (1971), 385–97; *Inventiones Math.* **11** (1970), 167–82.
‡ Ph.D. Thesis, Cambridge, 1974.

non-vanishing linear combination of $\omega_1, \omega_2, \eta_1, \eta_2$ and $2\pi i$; the auxiliary function now takes the form

$$\Phi(z_1, z_2, z_3) = \sum_{\lambda_0=0}^{L} \cdots \sum_{\lambda_3=0}^{L} p(\lambda_0, \ldots, \lambda_3)$$
$$\times (f(z_1, z_2, z_3))^{\lambda_0} (\wp_1(\omega_1 z_1))^{\lambda_1} (\wp_2(\omega_2 z_2))^{\lambda_2} e^{2\pi i \lambda_3 z_3},$$

where $L = [k^{\frac{1}{4}}]$ and $f(z_1, z_2, z_3)$ is the sum of $f(z_1, z_2)$, as defined above, and an algebraic multiple of πz_3. Here, however, it is necessary to appeal to another remarkable property of the division values, namely that, for any positive integer n, the field obtained by adjoining $\wp(\omega_1/n)$, $\wp(\omega_2/n)$, $\wp'(\omega_1/n)$ and $\wp'(\omega_2/n)$ to $K = Q(g_2, g_3, e^{2\pi i/n})$ has degree at most $2n^3$ over K; this ensures that the estimate $c'q^4$ referred to above remains unaltered in the present context. To complete the proof of Theorem 6.7 one has to establish the linear independence over the algebraic numbers of ω_1, η_1 and $2\pi i$ in the case when $\wp$ admits complex multiplication, and of these, together with ω_2, η_2, in the case when $\wp$ does not. The work runs on similar lines, using slightly modified auxiliary functions, but the determinant arguments at the end are no longer applicable; *ad hoc* techniques have been introduced to overcome this difficulty involving, in particular, new considerations on the density of zeros of meromorphic functions. The linear independence of ω_1, ω_2 and $2\pi i$ was in fact proved first by Coates utilizing a deep result of Serre, but Masser later verified this more elementarily.

In another direction, the work has been refined to yield estimates from below for linear forms in periods and quasi-periods. They show, for instance, that for any $\wp$-function with algebraic invariants, for any $\epsilon > 0$, and for any positive integer n,

$$|\wp(n)| < Cn^{(\log\log n)^{7+\epsilon}},$$

where C depends only on g_2, g_3 and ϵ.† In fact a similar result has been established for $\wp(\pi + n)$ and for $\wp(\alpha)$, where α is any non-zero algebraic number. The estimate compares well with the lower bound $|\wp(n)| > Cn$ valid for some $C > 0$ and infinitely many n.

Finally, as a further example of the type of theorem that has been obtained by the above methods, we mention a recent result of Masser‡ concerning algebraic points on elliptic curves; he has proved, namely, that if $\wp(z)$ has algebraic invariants and admits complex multiplication, then any numbers $u_1, \ldots, u_n$ for which $\wp(u_i)$ is algebraic are

† *Amer. J. Math.* **92** (1970), 619–22 (A. Baker); *P.C.P.S.* **73** (1973), 339–50 (D. W. Masser).
‡ Ph.D. Thesis, Cambridge, 1974.

either linearly dependent over $Q(\omega_1/\omega_2)$ or linearly independent over the field of all algebraic numbers. It would be of much interest to establish a theorem of the latter kind more generally for all $\wp$-functions with algebraic invariants, and it would likewise be of interest to extend Theorem 6.6 to apply to any number of $\wp$-functions; both problems, however, seem out of reach at present.

7

RATIONAL APPROXIMATIONS TO ALGEBRAIC NUMBERS

1. Introduction

In 1909, a remarkable improvement on Liouville's theorem was obtained by the Norwegian mathematician Axel Thue.† He proved that for any algebraic number α with degree $n > 1$ and for any $\kappa > \frac{1}{2}n+1$ there exists $c = c(\alpha, \kappa) > 0$ such that $|\alpha - p/q| > c/q^{\kappa}$ for all rationals p/q $(q > 0)$. His work rested on the construction of an auxiliary polynomial in two variables possessing zeros to a high order, and it can be regarded as the source of many of our modern transcendence techniques. The condition on κ was relaxed by Siegel‡ in 1921 to $\kappa > s + n/(s+1)$ for any positive integer s, thus, in particular, to $\kappa > 2\sqrt{n}$, and it was further relaxed by Dyson§ and Gelfond‖ independently in 1947 to $\kappa > \sqrt{(2n)}$. The latter expositions continued to involve polynomials in two variables and further progress seemed to require some extension of the arguments relating to polynomials in many variables; in fact special results in this connexion had already been obtained by Schneider¶ in 1936. A generalization of the desired kind was discovered by Roth†† in 1955; he showed indeed that the above proposition holds for any $\kappa > 2$, a condition which, in view of the introductory remarks of Chapter 1, is essentially best possible.

Roth's work, however, gave rise to a number of further problems. Siegel had initiated studies on the approximation of algebraic numbers by algebraic numbers in a fixed field, and also by algebraic numbers with bounded degree, and although Roth's arguments could be readily generalized to furnish a best possible result in connexion with the first topic,‡‡ they did not seem to admit a similar extension in connexion with the second. Even less, therefore, did they appear capable of dealing with the wider question concerning the simultaneous approximation of algebraic numbers by rationals. The whole subject was resolved by Schmidt§§ in 1970; building upon Roth's foundations but

† *J.M.* **135** (1909), 284–305.

‡ *M.Z.* **10** (1921), 173–213.

§ *Acta Math.* **79** (1947), 225–40.

‖ Bibliography.

¶ *J. M.* **175** (1936), 182–92.

†† *Mathematika*, **2** (1955), 1–20.

‡‡ See LeVeque (Bibliography).

§§ Bibliography.

introducing several new ideas, in particular from the Geometry of Numbers, he proved:

Theorem 7.1. *For any algebraic numbers* $\alpha_1, \ldots, \alpha_n$ *with* $1, \alpha_1, \ldots, \alpha_n$ *linearly independent over the rationals, and for any* $\epsilon > 0$, *there are only finitely many positive integers* q *such that*

$$q^{1+\epsilon}\|q\alpha_1\| \ldots \|q\alpha_n\| < 1.$$

Here $\|x\|$ denotes the distance of x from the nearest integer taken positively. The theorem implies, by a classical transference principle,† that there are only finitely many non-zero integers $q_1, \ldots, q_n$ with

$$|q_1 \ldots q_n|^{1+\epsilon} \|q_1\alpha_1 + \ldots + q_n\alpha_n\| < 1.$$

Further, as immediate corollaries, we see that there are only finitely many integers $p_1, \ldots, p_n, q$ $(q > 0)$ satisfying

$$|\alpha_j - p_j/q| < q^{-1-(1/n)-\epsilon} \quad (1 \leqslant j \leqslant n),$$

and also only finitely many integers $p, q_1, \ldots, q_n$ satisfying

$$|q_1\alpha_1 + \ldots + q_n\alpha_n - p| < q^{-n-\epsilon},$$

where $q = \max|q_j|$. Furthermore we have:

Theorem 7.2. *For any algebraic number* α *with degree exceeding* n *and any* $\epsilon > 0$, *there are only finitely many algebraic numbers* β *with degree at most* n *such that* $|\alpha - \beta| < B^{-n-1-\epsilon}$, *where* B *denotes the height of* β.

The theorem follows from the inequality just above with $\alpha_j = \alpha^j$, on noting that, if $P(x)$ is the minimal polynomial for β, then

$$|P(\alpha)| < BC|\alpha - \beta|$$

for some C depending only on α. The exponent of B is essentially best possible, as has been demonstrated by Wirsing.‡ In fact, Wirsing obtained Theorem 7.2 in 1965 before the work of Schmidt, but with the less precise exponent $-2n-\epsilon$.§

One of the main applications of the methods of this chapter has concerned Diophantine equations of norm form in several variables, which generalize the Thue equation discussed in Chapter 4; indeed the

† See Cassels' *Diophantine approximation* (Bibliography).
‡ *J. M.* **206** (1961), 67–77.
§ *Proc. Symposia Pure Math.* (*Amer. Math. Soc.*), **20** (1971), 213–47.

work has led to a complete description of all such equations that possess only finitely many solutions.†

Theorem 7.3. *Let K be an algebraic number field and let M be a module in K. A necessary and sufficient condition for there to exist an integer m such that the equation $N\mu = m$ has infinitely many solutions μ in M is that M be a full module in some subfield of K which is neither the rational nor an imaginary quadratic field.*

The necessity follows at once from the fact that the subfield, if it exists, contains at least one fundamental unit, and the sufficiency is a consequence of a generalized version of Theorem 7.1 relating to products of linear forms;‡ it is in fact a direct corollary in the case when the dimension of M is small compared with the degree of K. As examples, one sees that the equation

$$N(x_1+x_2\sqrt{2}+x_3\sqrt{3}) = 1$$

has infinitely many solutions in integers x_1, x_2, x_3 given by

$$x_1+x_2\sqrt{2} = \pm(1+\sqrt{2})^n, \quad \text{and by} \quad x_1+x_3\sqrt{3} = \pm(2+\sqrt{3})^n,$$

where $n = 0, 1, 2, \ldots$; and the equation

$$N(x_1+q^{1/p}x_2+\ldots+q^{(p-2)/p}x_{p-1}) = m,$$

where p, q are primes and m is any integer, has only a finite number of solutions in integers $x_1, \ldots, x_{p-1}$; for clearly the field generated by $q^{1/p}$ over the rationals has only trivial subfields. It should be noted, however, that, in contrast to the work of Chapter 4, the arguments here are not effective and cannot lead to a determination of the totality of solutions. In fact, apart from a few special results of Skolem,§ the only effective theorems established to date on equations of norm form in three or more variables derive from the work on the hypergeometric function referred to in § 5 of Chapter 4.‖

A generalization of Roth's theorem in the p-adic domain was obtained by Ridout¶ in 1957; in particular he proved that for any algebraic number α and any $\epsilon > 0$, there exist only finitely many integers p, q, comprised solely of powers of fixed sets of primes, such that $|\alpha - p/q| < q^{-\epsilon}$. In this case, however, Theorem 3.1 gives rather more; in fact, on taking $\alpha_1 = \alpha$ and the remaining α's as the given

† *M.A.* **191** (1971), 1–20.

‡ For an account of this and associated topics one may refer to the excellent survey of Schmidt; *Enseignement Math.* **17** (1971), 187–253.

§ Bibliography.

‖ *P.C.P.S.* **63** (1967), 693–702.

¶ *Mathematika*, **4** (1957), 125–31; **5** (1958), 40–8; see also Mahler (Bibliography).

primes, one sees at once that $q^{-\epsilon}$ can be replaced by $(\log q)^{-c}$ for some c depending only on α and the primes, a result moreover that is fully effective. Further theorems in the context of p-adic approximations follow from the other inequalities for $|\Lambda|$ recorded in Chapter 3.

2. Wronskians

The Wronskian of polynomials $\phi_1(x), \ldots, \phi_k(x)$ of one variable is defined as the determinant of order k with $(j!)^{-1}\phi_i^{(j)}(x)$ in the ith row and $(j+1)$th column, where $1 \leqslant i \leqslant k$, $0 \leqslant j < k$, and $\phi^{(j)}$ denotes the jth derivative of ϕ. Such Wronskians occurred in the original work of Thue, and they sufficed for the expositions of Siegel, Dyson and Gelfond; the arguments of Roth and Schmidt, however, involved the concept of a generalized Wronskian. Suppose that $\phi_1, \ldots, \phi_k$ are polynomials in n variables $x_1, \ldots, x_n$ and let $\Delta^{(j)}$ denote a differential operator of the form

$$(j_1! \ldots j_n!)^{-1} (\partial/\partial x_1)^{j_1} \ldots (\partial/\partial x_n)^{j_n},$$

where $j_1 + \ldots + j_n = j$. Then any determinant of order k with some $\Delta^{(j)}\phi_i$ in the ith row and $(j+1)$th column is called a generalized Wronskian of $\phi_1, \ldots, \phi_k$. There are clearly only finitely many generalized Wronskians of $\phi_1, \ldots, \phi_k$, and when $n = 1$ the set reduces to the original Wronskian. We shall require later the result that if $\phi_1, \ldots, \phi_k$ are linearly independent over their field of coefficients then some generalized Wronskian does not vanish identically; proofs are given, for instance, in the tracts of Cassels and Mahler.

3. The index

The proof of Theorem 7.1 involves polynomials P in kn variables x_{lm} $(1 \leqslant l \leqslant k, 1 \leqslant m \leqslant n)$, homogeneous in $x_{1m}, \ldots, x_{km}$ for each m. Suppose that P has real coefficients and let L_m $(1 \leqslant m \leqslant n)$ be real linear forms in $x_{1m}, \ldots, x_{km}$. Then the index of P with respect to $L_1, \ldots, L_n$ and positive integers $r_1, \ldots, r_n$ is defined as the largest value of θ such that P can be expressed as a linear combination of the polynomials $L_1^{j_1} \ldots L_n^{j_n}$ with

$$(j_1/r_1) + \ldots + (j_n/r_n) \geq \theta,$$

and with real polynomials in the x_{lm} as coefficients. It is easily verified that, for any polynomials

P, Q as above, the index, for brevity, ind, with respect to the L_m and r_m satisfies

$$\operatorname{ind}(P+Q) \geqslant \min(\operatorname{ind} P, \operatorname{ind} Q),$$
$$\operatorname{ind} PQ = \operatorname{ind} P + \operatorname{ind} Q.$$

We shall require also the related concept of the index of a real polynomial $P(x_1, \ldots, x_n)$ with respect to rationals p_m/q_m ($q_m > 0$) and integers $r_m > 0$ ($1 \leqslant m \leqslant n$); this is defined as the index of the polynomial

$$x_{21}^{d_1} \ldots x_{2n}^{d_n} P(x_{11}/x_{21}, \ldots, x_{1n}/x_{2n})$$

in the $2n$ variables x_{lm} ($l = 1, 2$) with respect to the linear forms

$$L_m = q_m x_{1m} - p_m x_{2m}$$

and the r_m, where d_m denotes the degree of P in x_m. The index in the latter sense occurred first in the work of Roth, and the generalized concept was introduced by Schmidt.

In analogy with the notation of earlier chapters, we define the height $\|P\|$ of a polynomial P as the maximum of the absolute values of its coefficients; we shall speak of the height only for polynomials with rational integer coefficients, not identically 0. The same definition will of course apply in the special case of linear forms.

Suppose now that P is a polynomial in kn variables as indicated at the beginning of the section. Let $L_1, \ldots, L_n$ be linear forms as there, with relatively prime integer coefficients, and let $q_m = \|L_m\|$. Further let $r_1, \ldots, r_n$ be positive integers such that $\delta r_m > r_{m+1}$ ($1 \leqslant m < n$), where $\delta = (\epsilon/32)^{2^n}$ and $0 < \epsilon < 1$. We have

Lemma 1. *If $q_m^{r_m} > q_1^{r_1}$ ($1 \leqslant m \leqslant n$) and $q_1^{\delta\eta} > 8^{nk^2}$, where $0 < \eta \leqslant k$, and if also P has height at most $q_1^{\delta\eta r_1/k^2}$ and degree at most r_m in $x_{1m}, \ldots, x_{km}$, then the index of P with respect to the L_m and r_m is at most ϵ.*

This is an extension, due to Schmidt, of the most fundamental part of Roth's work, sometimes called Roth's lemma. The result follows easily in fact from the case considered by Roth, as we now show.

We assume, as we may without loss of generality, that $q_m = |a_{1m}|$, where

$$L_m = \sum_{l=1}^{k} a_{lm} x_{lm} \quad (1 \leqslant m \leqslant n).$$

We shall further assume that (a_{1m}, a_{2m})† is at most $q_m^{(k-2)/(k-1)}$; this also involves no loss of generality, since a prime p can divide at most

† (a, b) denotes the greatest common divisor of a, b.

$k-2$ of the integers (a_{1m}, a_{lm}) with $1 < l \leqslant k$, whence their product divides q_m^{k-2}. Let now P' be the polynomial obtained from P by successively removing, in some order, the highest power of

$$x_{lm} \quad (3 \leqslant l \leqslant k, 1 \leqslant m \leqslant n)$$

that divides P and then setting the variable to 0; further let P'' be the polynomial obtained by setting $x_{2m} = 1$ in P' for each m. Then clearly the index of P with respect to the L_m and r_m is at most the index of P'' with respect to $-a_{2m}/a_{1m}$ and r_m. Also, by assumption, the denominator of a_{2m}/a_{1m}, when expressed in lowest terms, namely $q_m/(a_{1m}, a_{2m})$, is at least $q_m^{1/k}$. Hence we see that it suffices to prove the following modified version of Lemma 1.

For any integers r_m $(1 \leqslant m \leqslant n)$ as above and any rationals

$$p_m/q_m \quad (q_m > 0)$$

in their lowest terms such that $q_m^{r_m} > q_1^{\eta r_1}$ and $q_1^{\delta n} > 8^n$, where $0 < \eta \leqslant 1$, the index with respect to the p_m/q_m and r_m of any polynomial $P(x_1, \ldots, x_n)$ with height at most $q_1^{\delta\eta r_1}$ and degree at most r_m in x_m is at most ϵ.

Proofs of this proposition, possibly in slightly adapted form, in particular with $\eta = 1$, are given in several of the texts cited in the Bibliography, and our exposition can therefore be relatively brief. The result plainly holds for $n = 1$, for if j_1 is the exponent to which $x_1 - p_1/q_1$ divides $P(x_1)$ then, by Gauss' lemma, we have

$$P(x_1) = (q_1 x_1 - p_1)^{j_1} Q(x_1),$$

where Q is a polynomial with integer coefficients; thus the leading coefficient of P is at least $q_1^{j_1}$, whence $j_1/r_1 < \delta\eta < \epsilon$, as required. We now assume the validity of the proposition with n replaced by $n-1$ and we proceed to establish the assertion for n $(\geqslant 2)$.

We begin by writing P in the form

$$\phi_0\psi_0 + \ldots + \phi_{s-1}\psi_{s-1},$$

where the ϕ's and ψ's are polynomials in the variables $x_1, \ldots, x_{n-1}$ and x_n respectively with rational coefficients, and we choose one such representation for which s $(\leqslant r_n + 1)$ is minimal. Then there exist Wronskians U', V' of the ϕ's and ψ's respectively which do not vanish identically, and clearly $W = U'V'$ can be expressed as a determinant of order s with

$$\Delta^{(j)}(i!)^{-1}(\partial/\partial x_n)^i P(x_1, \ldots, x_n)$$

in the $(i+1)$th row and $(j+1)$th column, where the $\Delta^{(j)}$ are operators as in § 2 with $j_n = 0$. Hence W is a polynomial with degree at most sr_j in x_j and with

$$\|W\| \leqslant (8^{rn}\|P\|)^s \leqslant q_1^{2\delta \eta rs},$$

where $r = r_1 = \max r_m$; here we are using the hypothesis $q_1^{\delta\eta} > 8^n$ and the observations that $\Delta^{(j)}$ acting on any monomial in P introduces a factor not exceeding 2^{rn}, that there are at most 2^{rn} such monomials, and that the number of terms obtained on expanding the determinant, for W is $s! \leqslant 2^{rs}$. Now, again by Gauss' lemma, we have $W = UV$, where U, V are polynomials with integer coefficients in the variables $x_1, \ldots, x_{n-1}$ and x_n respectively, given by some rational multiples of U', V'; and clearly the bound for $\|W\|$ obtains also for $\|U\|$ and $\|V\|$. Thus, by our inductive hypothesis, it follows, on taking 2δ in place of δ, that the index of U with respect to the p_m/q_m and r_m is at most $2^{-5+1/2^{n-1}}s\epsilon^2$. Further, by the case $n = 1$ of the proposition together with the hypothesis $q_n^{r_n} \geqslant q_1^{\eta r_1}$, the same bound applies for the index of V. We conclude therefore that the index of W is at most $\frac{1}{8}s\epsilon^2$.

On the other hand, the index of the general element in the determinant for W is at least

$$\phi_i - \sum_{m=1}^{n-1} j_m/r_m,$$

where $\phi_i = \theta - i/r_n$, θ denotes the index of P, and

$$j_1 + \ldots + j_{n-1} = j \leqslant s-1 \leqslant r_n;$$

further, by hypothesis, we have $\delta r_m > r_{m+1}$ and so the above sum is at most δ. Hence the index of W is at least

$$\sum_{i=0}^{s-1} \max(\phi_i - \delta, 0) \geqslant \sum_{i=0}^{s-1} \max(\phi_i, 0) - s\delta.$$

But if $\theta r_n < s-1$ then the last sum is

$$([\theta r_n]+1)(\theta - [\theta r_n]/(2r_n)) \geqslant \tfrac{1}{4}\theta^2 s,$$

and if $\theta r_n \geqslant s-1$ then it is

$$\theta s - \tfrac{1}{2}s(s-1)/r_n \geqslant \tfrac{1}{2}\theta s.$$

On comparing estimates, we obtain

$$\max(\tfrac{1}{2}\theta, \tfrac{1}{4}\theta^2) \leqslant \tfrac{1}{8}\epsilon^2 + \delta \leqslant \tfrac{1}{4}\epsilon^2,$$

whence $\theta \leqslant \epsilon$, as required.

4. A combinatorial lemma

We prove now a lemma of a combinatorial nature relating to the law of large numbers.† A result of this kind occurred first in the work of Schneider, and it was utilized later by Roth who gave a simplified proof due to Davenport. Another proof, attributed to Reuter, and furnishing a slightly stronger theorem, was given by Mahler in his tract, and Schmidt subsequently obtained the generalization we establish here.

Lemma 2. *Suppose that $r_1, \ldots, r_n$ and k are positive integers and that $0 < \epsilon < 1$. Then the number of non-negative integers*

$$j_{lm} \quad (1 \leqslant l \leqslant k,\ 1 \leqslant m \leqslant n)$$

satisfying

$$\sum_{l=1}^{k} j_{lm} = r_m \quad (1 \leqslant m \leqslant n), \qquad \sum_{m=1}^{n} j_{1m}/r_m < n/k - \epsilon n,$$

is at most
$$\binom{r_1+k-1}{r_1} \cdots \binom{r_n+k-1}{r_n} e^{-\frac{1}{4}\epsilon^2 n}.$$

We commence the proof by observing that the required number N of integers j_{lm} is given by
$$\Sigma \nu_1(j_{11}) \ldots \nu_n(j_{1n}),$$

where the sum is over all non-negative integers $j_{11}, \ldots, j_{1n}$ satisfying the given inequality, and $\nu_m(j)$ denotes the number of solutions of the equation
$$\sum_{l=2}^{k} j_{lm} = r_m - j$$

in non-negative integers $j_{2m}, \ldots, j_{km}$, that is

$$\nu_m(j) = \binom{r_m - j + k - 2}{k-2}.$$

Hence we see that the multiple sum

$$\sum_{j_1=0}^{r_1} \cdots \sum_{j_m=0}^{r_m} \nu_1(j_1) \ldots \nu_m(j_m) \exp\left\{\tfrac{1}{2}\epsilon\left(n/k - \sum_{m=1}^{n} j_m/r_m\right)\right\}$$

is at least $N e^{\frac{1}{4}\epsilon^2 n}$. Now the sum can be written alternatively in the form

$$\prod_{m=1}^{n} \left\{\sum_{j_m=0}^{r_m} \nu_m(j_m) \exp\left(\tfrac{1}{2}\epsilon \rho_m\right)\right\},$$

† Cf. the paper of Wirsing cited earlier: *Proc. Symposia Pure Math.* (*Amer. Math. Soc.*), **20** (1971), 213–47.

where $\rho_m = 1/k - j_m/r_m$, and clearly $|\rho_m| \leqslant 1$. But if $|x| \leqslant 1$ then $e^x < 1+x+x^2$, and so

$$\exp(\tfrac{1}{2}\epsilon\rho_m) < \tfrac{1}{2}\epsilon\rho_m + \exp(\tfrac{1}{4}\epsilon^2).$$

Further we have
$$\sum_{j_m=0}^{r_m} \nu_m(j_m)\rho_m = 0;$$

for ρ_m can plainly be expressed as

$$(r_m - j_m)/r_m - (1-1/k),$$

and it is easily verified by induction on r that

$$\sum_{j=0}^{r} \binom{r-j+k-2}{k-2} = \binom{r+k-1}{r},$$

$$\sum_{j=0}^{r} j\binom{j+k-2}{k-2} = r\left(1-\frac{1}{k}\right)\binom{r+k-1}{r}.$$

Thus, on appealing again to the first of the above binomial identities, we obtain
$$\prod_{m=1}^{n}\left\{\binom{r_m+k-1}{r_m} e^{\frac{1}{4}\epsilon^2}\right\} \geqslant N e^{\frac{1}{4}\epsilon^2 n},$$

and this gives the asserted estimate.

5. Grids

Let T be a subspace of k-dimensional Euclidean space spanned by linearly independent vectors $\mathbf{u}_1, \ldots, \mathbf{u}_{k-1}$. By a grid of size s on T we shall mean the finite set of vectors of the form

$$w_1\mathbf{u}_1 + \ldots + w_{k-1}\mathbf{u}_{k-1},$$

where $w_1, \ldots, w_{k-1}$ run through all rational integers with $1 \leqslant w_l \leqslant s$.

Now let T_m $(1 \leqslant m \leqslant n)$ be any subspaces as above, and let Γ_m be a grid of size s_m on T_m. Further let T, Γ signify the cartesian products $T_1 \times \ldots \times T_n$ and $\Gamma_1 \times \ldots \times \Gamma_n$ respectively. We shall denote by P a polynomial as indicated at the beginning of § 3 with degree r_m in $x_{1m}, \ldots, x_{km}$, and we shall signify by $\Delta_m^{(j)}$ a differential operator as in § 2, acting on $x_{1m}, \ldots, x_{km}$. The following simple lemma, due to Schmidt, is fundamental to the proof of Theorem 7.1.

Lemma 3. *If, for some integers t_m $(1 \leqslant m \leqslant n)$ with $s_m(t_m+1) > r_m$, all polynomials $\Delta_1^{(j_1)} \ldots \Delta_n^{(j_n)} P$ with $j_m \leqslant t_m$ vanish everywhere on Γ, then P vanishes identically on T.*

It is clear that the lemma follows at once by induction from the case $n = 1$, and it will suffice therefore to prove the latter. Further, one can obviously assume, by applying a linear transformation, that T_m is the plane $x_{km} = 0$, with basis consisting of the first $k-1$ rows of the unit matrix of order k. Thus, omitting the suffix m, we see that it is enough to prove:

A polynomial $P(x_1, \ldots, x_{k-1})$ with degree r vanishes identically if all $\Delta^{(j)}P$ with $j \leqslant t$ vanish at all integer points $(w_1, \ldots, w_{k-1})$ with $1 \leqslant w_l \leqslant s$, where $s(t+1) > r$.

Here $\Delta^{(j)}$ denotes a differential operator on $x_1, \ldots, x_{k-1}$ of order j. The assertion is clearly valid for $k = 2$, since a polynomial in one variable with degree r cannot have more than r zeros, and we shall assume the proposition when k is replaced by $k-1$. If now P does not vanish identically then there is a largest integer q such that the rational function

$$Q = (x_1 - 1)^{-q} \ldots (x_1 - s)^{-q} P$$

is in fact a polynomial, and since, by hypothesis, $s(t+1) > r$, we have $q \leqslant t$. Further, by choice of q, one at least of the polynomials $Q(w_1, x_2, \ldots, x_{k-1})$ with $1 \leqslant w_1 \leqslant s$ does not vanish identically; let this be R. Then $\Delta^{(j)}R$ vanishes at all integer points $(w_2, \ldots, w_{k-1})$ with $1 \leqslant w_l \leqslant s$, where $\Delta^{(j)}$ is any differential operator on $x_2, \ldots, x_{k-1}$ with order $j \leqslant t-q$. But R has degree at most $r - sq < (t-q+1)\,s$, and this is plainly contrary to the inductive hypothesis. The contradiction establishes the assertion.

6. The auxiliary polynomial

For each m with $1 \leqslant m \leqslant n$ we shall denote by L_{lm} $(1 \leqslant l \leqslant k)$ linear forms in $x_{1m}, \ldots, x_{km}$ with real algebraic integer coefficients. Further we shall denote by d the degree of the field K generated by all the coefficients over the rationals, and we shall signify by $c_1, c_2, \ldots$ numbers greater than 1 which depend on these coefficients only.

Let now $r_1, \ldots, r_n$ be any positive integers, and let $r = \max r_m$. Further suppose that $0 < \epsilon < 1$ and that $e^{\frac{1}{4}\epsilon^2 n} > 2kd$. Adopting the notation of § 3, we have

Lemma 4. *There is a polynomial P with degree at most r_m in $x_{1m}, \ldots, x_{km}$ and with height at most c_1^r such that, for each l with $1 \leqslant l \leqslant k$, the index of P with respect to the L_{lm} and r_m is at least $n/k - \epsilon n$.*

It can be assumed, without loss of generality, that, for all l, m, the coefficient of x_{1m} in L_{lm}, say α_{lm}, is not 0. Then P has to be determined such that, for all l and all non-negative integers $j_1, \ldots, j_n$ with

$$\sum_{m=1}^{n} j_m/r_m < n/k - \epsilon n,$$

the polynomials

$$(j_1! \ldots j_n!)^{-1} (\partial/\partial x_{11})^{j_1} \ldots (\partial/\partial x_{1n})^{j_n} P$$

vanish identically when $-L_{lm}$, with x_{1m} equated to 0, is substituted for x_{1m}, and the factor α_{lm} is included to multiply each of $x_{2m}, \ldots, x_{km}$. Now these polynomials are homogeneous in $x_{2m}, \ldots, x_{km}$ with degree $r_m - j_m$ and hence, by Lemma 2, they have, in total, at most $kN e^{-\frac{1}{4}\epsilon^2 n}$ coefficients, where N denotes the product of binomial factors occurring in the enunciation of the lemma. Each coefficient is a linear form in the coefficients of P, and there are precisely N of the latter. Furthermore, the coefficients in the linear forms are algebraic integers in K with sizes at most c_2^r (cf. the estimates in § 3). It follows, on utilizing an integral basis for K and recalling the hypothesis $e^{\frac{1}{4}\epsilon^2 n} > 2kd$, that one has to satisfy at most $\frac{1}{2}N$ linear equations with rational integer coefficients each having absolute value at most c_3^r (cf. § 3, Chapter 6). The required result is now obtained from Lemma 1 of Chapter 2.

7. Successive minima

We recall from the Geometry of Numbers that if R is any convex body in k-dimensional Euclidean space, then the numbers λ_l $(1 \leqslant l \leqslant k)$, given by the infimum of all $\lambda > 0$ such that λR contains l linearly independent integer points, are termed the successive minima of R, and they have the property that $\lambda_1 \ldots \lambda_k V$, where V denotes the volume of R, is bounded above and below by positive numbers depending only on k.

We now combine the preceding lemmas to obtain a proposition on the penultimate minimum of a certain parallelepiped, which will be the main instrument in the proof of Theorem 7.1. We shall denote by $M_1, \ldots, M_k$ linear forms in $x_1, \ldots, x_k$ with real algebraic integer coefficients constituting a non-singular matrix $\mathbf{A}$, and we shall signify by $M'_1, \ldots, M'_k$ the adjoint linear forms with coefficients given by the columns of $\mathbf{A}^{-1}$. Further we shall signify by S some non-empty set of suffixes i such that M'_i does not represent zero for any integral values, not all 0, of the variables; the assumption that S exists involves, of course, some loss of generality. We prove:

Lemma 5. *For any $\zeta > 0$ there exists $c > 0$ such that for all positive $\mu_1, \ldots, \mu_k$ satisfying $\mu_1 \ldots \mu_k = 1$ and $\mu_i \geqslant 1$ for i in S, the penultimate minimum λ_{k-1} of the parallelepiped $|M_l| \leqslant \mu_l$ $(1 \leqslant l \leqslant k)$ exceeds $\mu^{-\zeta}$, where μ denotes the maximum of $\mu_1, \ldots, \mu_k$ and c.*

It will be seen that the lemma immediately implies Roth's theorem, that is the case $n = 1$ of Theorem 7.1; this follows on taking

$$M_1 = \alpha_1 x_1 - x_2, \quad M_2 = x_2$$

and S to consist just of the suffix 2, as is possible since α_1 is irrational.

We show first that it suffices to prove a modified version of Lemma 5. Suppose that $Q \geqslant \mu^k$ and let $\omega_1, \ldots, \omega_k$ be defined by $\mu_l = Q^{\omega_l}$. Then since $\mu_1 \ldots \mu_k = 1$ we have $\omega_1 + \ldots + \omega_k = 0$ and clearly $\omega_i \geqslant 0$ for i in S. Clearly also $\omega_l \leqslant 1$ for all l and, since again $\mu_1 \ldots \mu_k = 1$ we have $\mu_l \geqslant Q^{-1}$, whence $\omega_l \geqslant -1$. Now for any positive integer N there are rationals $\omega_1', \ldots, \omega_n'$ with denominator N satisfying $|\omega_l - \omega_l'| < 1/N$ and $|\omega_l'| \leqslant 1$ for all l, and also $\omega_1' + \ldots + \omega_k' = 0$; indeed one has merely to take $N\omega_1' = [N\omega_1]$ and, having defined $\omega_1', \ldots, \omega_{l-1}'$, to take $N\omega_l'$ as $[N\omega_l]$ or $-[-N\omega_l]$ according as $\omega_1' + \ldots + \omega_{l-1}'$ does or does not exceed $\omega_1 + \ldots + \omega_{l-1}$. Plainly the $\omega_1', \ldots, \omega_k'$ belong to a finite set of rationals independent of Q, and the minimum λ_{k-1}' of the parallelepiped $|M_l| \leqslant Q^{\omega_l'}$ $(1 \leqslant l \leqslant k)$ exceeds $Q^{-1/N}\lambda_{k-1}$. Hence it is enough to prove:

For any real $\omega_1, \ldots, \omega_k$ with $\omega_1 + \ldots + \omega_k = 0$, $|\omega_l| \leqslant 1$ $(1 \leqslant l \leqslant k)$ and $\omega_i \geqslant 0$ for all i in S, and for any $\zeta > 0$, there exists $C > 0$ such that, for all $Q > C$, the minimum λ_{k-1} of the parallelepiped

$$|M_l| \leqslant Q^{\omega_l} \quad (1 \leqslant l \leqslant k)$$

exceeds $Q^{-\zeta}$.

We shall suppose that $\zeta \leqslant 1$, as obviously we may, and we shall signify by d the degree of the field generated by the elements of **A** over the rationals. Let ϵ be any positive number less than $\zeta/(8k)^2$, let n be any integer satisfying the condition preceding Lemma 4, and let δ be defined as in Lemma 1. We shall assume that there is an unbounded set of values of Q such that $\lambda_{k-1} \leqslant Q^{-\zeta}$, and we shall ultimately derive a contradiction. We select a sufficiently large Q_1 in this set, that is $Q_1 > c_1$, where c_1, like c_2, c_3 below, depends only on **A**, $k, n, d, \epsilon, \delta, \zeta$ and the ω's. We then select further elements $Q_2, \ldots, Q_n$ in the set such that $Q_m^{\frac{1}{2}\delta} > Q_{m-1}$ $(1 < m \leqslant n)$, whence clearly $Q_1 < \ldots < Q_n$. Finally we choose positive integers $r_1, \ldots, r_n$ such that $Q_1^{\epsilon r_1} > Q_n$ and

$$Q_1^{r_1} \leqslant Q_m^{r_m} \leqslant Q_1^{(1+\epsilon)r_1} \quad (1 \leqslant m \leqslant n);$$

then plainly the condition preceding Lemma 1 is satisfied.

We observe now that the hypotheses of Lemma 4 hold when $L_{lm} = M_l(\mathbf{x}_m)$, where $\mathbf{x}_m$ denotes the vector $(x_{1m}, \ldots, x_{km})$; let P be the polynomial constructed there. Further we note that, for any Q as above, there exist linearly independent integer points $\mathbf{u}_1, \ldots, \mathbf{u}_k$ with $\mathbf{u}_l$ in $\lambda_l R$, where R denotes the given parallelepiped and $\lambda_1, \ldots, \lambda_k$ its successive minima. Moreover, there is a linear form L with relatively prime integer coefficients, unique except for a factor ± 1, which vanishes at $\mathbf{u}_1, \ldots, \mathbf{u}_{k-1}$; we take $\mathbf{u}_{lm}$ and L_m to be these $\mathbf{u}_l$ and L respectively when $Q = Q_m$. We shall verify later that, if Q is sufficiently large, then $q = \|L\|$ satisfies $Q^c \leqslant q \leqslant Q^{c'}$, where c, c' are positive numbers depending only on ζ and d. Assuming this for the present, it follows that all the hypotheses of Lemma 1 are satisfied with $\eta = c/c'$, provided that c_1, and so also q_1 and Q_1, are large enough. Hence we conclude that the index of P with respect to the L_m and r_m is at most ϵ.

We proceed to prove that, with the notation of § 5, all polynomials ΔP with

$$\Delta = \Delta_1^{(j_1)} \ldots \Delta_n^{(j_n)}, \quad \sum_{m=1}^{n} j_m/r_m < 2\epsilon n$$

vanish everywhere on Γ, where Γ_m is the grid of size $[\epsilon^{-1}]+1$ on the space T_m spanned by $\mathbf{u}_{lm}$ $(1 \leqslant l < k)$. This implies, by Lemma 3, on taking $t_m = [\epsilon r_m]$, that all polynomials ΔP, with $\Sigma j_m/r_m < \epsilon n$, vanish identically on the $n(k-1)$-dimensional space of solutions of

$$L_1 = \ldots = L_n = 0.$$

But the latter contradicts the above conclusion concerning the index of P, and the contradiction establishes the lemma. To prove the proposition, let ΔP be any of the polynomials in question and let P' be the polynomial in new variables y_{lm} obtained from ΔP by the linear substitution $y_{lm} = L_{lm}$. Then it is readily verified that P' has height at most c_2^r, where $r = \max r_m$. Further, since, by assumption, $\lambda_{k-1} \leqslant Q^{-\zeta}$, we have for any $\mathbf{x}_m$ on Γ_m

$$|y_{lm}| < k(\epsilon^{-1}+1)\, Q_m^{\omega_l - \zeta} < Q_m^{\omega_l - \frac{1}{2}\zeta}.$$

Thus, by Lemma 4, it follows that, for all points on Γ, we have $|\Delta P| < c_3^r e^s$, where

$$s = \sum_{l=1}^{k} \sum_{m=1}^{n} (\omega_l - \tfrac{1}{2}\zeta) j_{lm} \log Q_m,$$

and j_{lm} are some non-negative integers with

$$\sum_{l=1}^{k} j_{lm} \leqslant r_m \quad (1 \leqslant m \leqslant n), \qquad n/k - \sum_{m=1}^{n} j_{lm}/r_m < 3\epsilon n \quad (1 \leqslant l \leqslant k).$$

Denoting, for brevity, the left-hand side of the last inequality by h_l, we see that, by the first inequality, $h_1 + \ldots + h_k \geqslant 0$, and so both inequalities together imply that $|h_l| < 3kn\epsilon$. Further, since $|\omega_l| \leqslant 1$, we obtain, in view of the initial choice of $r_1, \ldots, r_n$,

$$s \leqslant r_1 \log Q_1 \sum_{l=1}^{k} \sum_{m=1}^{n} \{(\omega_l - \tfrac{1}{2}\zeta) j_{lm}/r_m + 2\epsilon\}.$$

But now, by virtue of our estimate for h_l and the hypothesis

$$\omega_1 + \ldots + \omega_k = 0,$$

the double sum here differs from $-\frac{1}{2}\zeta n$ by at most $8k^2n\epsilon$. Since, by definition, $\epsilon < \zeta/(8k)^2$, it follows that $|\Delta P| < Q_1^{-\frac{1}{4}\zeta nr} < 1$, provided Q_1 is sufficiently large. On the other hand, ΔP is a rational integer for all points on Γ, and hence $\Delta P = 0$, as required.

It remains only to prove the assertion concerning $q = \|L\|$. Let $\mathbf{U}$ be the matrix with columns $\mathbf{u}_1, \ldots, \mathbf{u}_k$ and let $\mathbf{v}_1, \ldots, \mathbf{v}_k$ be the rows of $\mathbf{U}^{-1}$. Then clearly $\rho\mathbf{v}_k$ is the coefficient vector of L for some rational ρ. Since $L(\mathbf{u}_k)$ is an integer and $\mathbf{v}_k\mathbf{u}_k = 1$, ρ is in fact an integer. Further ρ divides $\det \mathbf{U}$, for plainly $\mathbf{U}^{-1} = \operatorname{adj} \mathbf{U}/\det \mathbf{U}$.† Furthermore we have $\det \mathbf{U} \ll 1$, where the implied constant depends only on $\mathbf{A}$,‡ for certainly R has volume $\gg 1$ and hence, by the property of successive minima quoted at the beginning, $\det(\mathbf{AU}) \ll 1$. It follows that each element of $(\det \mathbf{U})\,\mathbf{v}_k$ is a rational integer $\ll q$. Hence the element in the kth row and lth column of $\operatorname{adj}(\mathbf{AU})$, namely $(\det(\mathbf{AU}))\, M_l'(\mathbf{v}_k)$, is an algebraic integer with size $\ll q$. But by hypothesis we have $\lambda_{k-1} < Q^{-\zeta}$ and $\omega_1 + \ldots + \omega_k = 0$, and thus the element is also $\ll Q^{-\omega_l-(k-1)\zeta}$. We conclude that, for l in S, the element is both $\gg q^{-d}$ and $\ll Q^{-(k-1)\zeta}$, and, since S is assumed non-empty, this gives the required lower bound for q. The upper bound follows from the identity $\mathbf{U}^{-1} = (\mathbf{AU})^{-1}\mathbf{A}$, on observing, as above, that the elements in the kth row of $(\mathbf{AU})^{-1}$ are $\ll Q$.

8. Comparison of minima

We prove first a general lemma of Davenport, and we proceed then to show that, with some proviso, the minima λ_{k-1} and λ_k of the parallelepiped of Lemma 5 differ only by a small factor. Constants implied by $\ll$ will depend only on k.

† 'det' and 'adj' are abbreviations for determinant and adjoint respectively.

‡ We are using Vinogradov's notation; by $a \ll b$ one means $|a| < bc$ for some constant c, and similarly for $\gg$.

Lemma 6. *Let $L_1, \ldots, L_k$ be real linear forms with determinant* 1 *and let $\lambda_1, \ldots, \lambda_k$ be the successive minima of the parallelepiped*

$$|L_l| \leqslant 1 \quad (1 \leqslant l \leqslant k).$$

Suppose that $\rho_1 \geqslant \ldots \geqslant \rho_k > 0$ and that

$$\rho_1\lambda_1 \leqslant \ldots \leqslant \rho_k\lambda_k, \quad \rho_1 \ldots \rho_k = 1.$$

Then for some permutation $\rho_1', \ldots, \rho_k'$ of $\rho_1, \ldots, \rho_k$, the successive minima $\lambda_1', \ldots, \lambda_k'$ of the parallelepiped $\rho_l'|L_l| \leqslant 1$ $(1 \leqslant l \leqslant k)$ satisfy

$$\rho_l\lambda_l \ll \lambda_l' \ll \rho_l\lambda_l \quad (1 \leqslant l \leqslant k).$$

Proof. There certainly exist linearly independent integer points $\mathbf{x}_1, \ldots, \mathbf{x}_k$ such that one at least of $|L_1|, \ldots, |L_k|$ assumes the value λ_l at $\mathbf{x}_l$, and we denote by S_l the space spanned by $\mathbf{x}_1, \ldots, \mathbf{x}_l$. Further, for each $l \geqslant 2$, there is a non-trivial linear relation $a_1 L_1 + \ldots + a_l L_l = 0$ satisfied identically on S_{l-1}, and $L_1, \ldots, L_k$ can be permuted so that $|a_l|$ is maximal; this gives

$$|L_1| + \ldots + |L_{l-1}| > \tfrac{1}{2}(|L_1| + \ldots + |L_l|)$$

identically on S_{l-1}, whence by induction

$$|L_1| + \ldots + |L_l| \geqslant 2^{l-k}(|L_1| + \ldots + |L_k|)$$

identically on S_l for $l = 1, 2, \ldots, k$. Now for any j it is clear that every point in S_j not in S_{j-1} satisfies

$$\max(|L_1|, \ldots, |L_k|) \geqslant \lambda_j,$$

and thus, in view of the inequality obtained above, it satisfies also

$$\max(\rho_1|L_1|, \ldots, \rho_k|L_k|) \gg \rho_j\lambda_j.$$

By hypothesis, $\rho_j\lambda_j \geqslant \rho_l\lambda_l$ for $j \geqslant l$, and the required lower bound for λ_l' follows on taking $\rho_1', \ldots, \rho_k'$ to be the permutation of $\rho_1, \ldots, \rho_k$ inverse to that associated with $L_1, \ldots, L_k$. The upper bound is a consequence of the equation $\rho_1 \ldots \rho_k = 1$ together with the property, noted earlier, that $\lambda_1 \ldots \lambda_k$ and $\lambda_1' \ldots \lambda_k'$ are both $\ll 1$ and $\gg 1$.

Lemma 7. *The last two minima of the parallelepiped of Lemma* 5 *satisfy $\lambda_{k-1} \gg \lambda_k \mu^{-k\zeta}$, provided that $\lambda_1\mu_i > \mu^{-\zeta}$ for all i in S.*

Proof. The hypotheses of Lemma 6 hold with L_l $(1 \leqslant l \leqslant k)$ given by $\mu_l^{-1}M_l$ and

$$\rho_l = \rho/\lambda_l \quad (1 \leqslant l < k), \quad \rho_k = \rho/\lambda_{k-1},$$

where ρ is defined by the equation $\rho_1 \dots \rho_k = 1$. Let $\rho'_1, \dots, \rho'_k$ be the permutation of $\rho_1, \dots, \rho_k$ indicated in the lemma, and let $\mu'_l = \mu_l/\rho'_l$. Assume first that $\mu'_i \geqslant 1$ for all i in S. Then from Lemma 5 with μ_l replaced by μ'_l, we infer that, for any $\zeta' > 0$, there exists $c' > 0$ such that $\lambda'_{k-1} > \mu'^{-\zeta'}$, where μ' denotes the maximum of $\mu'_1, \dots, \mu'_k$ and c'. On the other hand, from Lemma 6, $\lambda'_{k-1} \ll \rho_{k-1}\lambda_{k-1} = \rho$, and clearly, since $\lambda_1 \dots \lambda_k \ll 1$, we have $\rho^k \ll \lambda_{k-1}/\lambda_k$. Thus it suffices to prove that $\mu'^{\zeta'} \ll \mu^{\zeta}$ if ζ' is chosen sufficiently small. But by hypothesis, since S is assumed non-empty, we have $\lambda_1\mu > \mu^{-\zeta}$; further, since $\lambda_l \geqslant \lambda_1$ for all l, we see that $\rho \gg \lambda_1$ and $\lambda_1^{k-1}\lambda_{k-1} \ll 1$. Hence we obtain

$$\mu' \ll \mu\lambda_{k-1}/\rho \ll \mu\lambda_1^{-k} < \mu^{k(\zeta+1)+1},$$

and the required result follows. If, contrary to the above assumption, $\mu'_i < 1$ for some i in S, then, on observing that by hypothesis

$$\rho\mu'_i \geqslant \lambda_1\mu_i > \mu^{-\zeta},$$

we obtain $\rho > \mu^{-\zeta}$ and the required result again follows.

9. Exterior algebra

For any vectors $\mathbf{x}_1, \dots, \mathbf{x}_l$ in R^k with $1 \leqslant l < k$, one denotes by $\mathbf{x}_1 \wedge \dots \wedge \mathbf{x}_l$ the vector in R^m whose elements are the $m = \binom{k}{l}$ subdeterminants of order l formed from the k by l matrix with columns $\mathbf{x}_1, \dots, \mathbf{x}_l$. We shall utilize some well-known properties of this product; in particular, we shall require Laplace's identity

$$(\mathbf{x}_1 \wedge \dots \wedge \mathbf{x}_l)(\mathbf{y}_1 \wedge \dots \wedge \mathbf{y}_l) = \det(\mathbf{x}_i\mathbf{y}_j),$$

where on the left one has the usual vector dot product, and also the relation

$$\det \mathbf{A}_\sigma = (\det \mathbf{A})^{lm/k},$$

which holds for any matrix $\mathbf{A}$ of order k with column vectors $\mathbf{a}_1, \dots, \mathbf{a}_k$, say, where $\mathbf{A}_\sigma = \mathbf{a}_{i_1} \wedge \dots \wedge \mathbf{a}_{i_l}$ and σ runs through all sets of l distinct integers $i_1, \dots, i_l$ from $1, \dots, k$.†

We shall need, in addition, the following lemma, due to Mahler, on compound convex bodies. $\mathbf{A}$ will signify a matrix as above with $\det \mathbf{A} = 1$, and, as in § 8, constants implied by $\ll$ will depend only on k. Further we shall denote by $\mathbf{ax}$ the linear form in the elements of $\mathbf{x}$ with coefficient vector $\mathbf{a}$.

† Short proofs are given in Schmidt's tract (Bibliography).

Lemma 8. *The successive minima* $\lambda_1, \ldots, \lambda_k$ *and* $\nu_1, \ldots, \nu_m$ *of the parallelepipeds* $|\mathbf{a}_i \mathbf{x}| \leqslant 1$ $(1 \leqslant i \leqslant k)$ *and* $|\mathbf{A}_\sigma \mathbf{X}| \leqslant 1$, *respectively, satisfy*

$$\lambda_{\tau_i} \ll \nu_i \ll \lambda_{\tau_i} \quad (1 \leqslant i \leqslant m),$$

where τ *runs through all sets* σ *as above,* $\lambda_\tau = \Pi \lambda_j$, *the product being taken over all* j *in* τ, *and* $\lambda_{\tau_1} \leqslant \lambda_{\tau_2} \leqslant \ldots \leqslant \lambda_{\tau_m}$.

Proof. Let $\mathbf{x}_1, \ldots, \mathbf{x}_k$ be linearly independent integer points such that $|\mathbf{a}_i \mathbf{x}_j| \leqslant \lambda_j$ $(1 \leqslant j \leqslant k)$, and let $\mathbf{X}_\tau$ be defined like $\mathbf{A}_\sigma$ above, with $\mathbf{x}$ in place of $\mathbf{a}$. By Laplace's identity we have

$$|\mathbf{A}_\sigma \mathbf{X}_\tau| = |\det(\mathbf{a}_i \mathbf{x}_j)| \leqslant m!\, \lambda_\tau,$$

where i, j run through all elements of σ, τ respectively. Hence, for each i with $1 \leqslant i \leqslant m$, we have $|\mathbf{A}_\sigma \mathbf{X}_{\tau_i}| \ll \lambda_{\tau_i}$ and so $\nu_i \ll \lambda_{\tau_i}$. But since, by hypothesis, $\det \mathbf{A} = 1$, we have $\det \mathbf{A}_\sigma = 1$, and thus the volume of the parallelepiped $|\mathbf{A}_\sigma \mathbf{X}| \leqslant 1$ is 2^m. Thus $\nu_1 \ldots \nu_m \gg 1$ and since $\lambda_{\tau_1} \ldots \lambda_{\tau_m} \ll 1$ it follows that $\nu_i \gg \lambda_{\tau_i}$, as required.

10. Proof of main theorem

It will suffice to prove Theorem 7.1 under the assumption that $\alpha_1, \ldots, \alpha_n$ are real algebraic integers, for clearly the general result then follows on multiplying each α_j by the leading coefficient in its minimal polynomial. We shall signify by $\mathbf{a}_j$ $(1 \leqslant j \leqslant n)$ the vector in R^{n+1} given by $(\mathbf{e}_j, \alpha_j)$, where $\mathbf{e}_1, \ldots, \mathbf{e}_n$ denote the rows of the unit matrix of order n. Further, for brevity, we shall write $k = n + 1$, and we shall denote by $\mathbf{a}_k$ the vector $(0, \ldots, 0, 1)$ in R^k. Constants implied by $\ll$ or $\gg$ will depend only on $\alpha_1, \ldots, \alpha_n, k, \epsilon$ and the quantities ξ, ζ to be defined below.

We show first that the theorem is a consequence of the following proposition.

For any $\xi > 0$ *and any positive numbers* $\mu_1, \ldots, \mu_k$ *with* $\mu_1 \ldots \mu_k = 1$ *and* $\mu_j < 1$ $(1 \leqslant j < k)$, *the first minimum* λ_1 *of the parallelepiped* $|\mathbf{a}_j \mathbf{x}| \leqslant \mu_j$ $(1 \leqslant j \leqslant k)$ *exceeds* $\mu^{-\xi}$ *if* $\mu \gg \mu_k$ *and* $\mu \gg 1$.

The proof proceeds by induction on n; we have already remarked that the case $n = 1$ is an immediate consequence of Lemma 5, and we assume now that the theorem holds when n is replaced by $n-1$. Let q be a positive integer satisfying the inequality occurring in the enunciation, and let

$$\mu_j = q^{\epsilon/(2n)} \|q\alpha_j\| \quad (1 \leqslant j \leqslant n).$$

Further let $\mu_k = (\mu_1 \dots \mu_n)^{-1}$, where $k = n+1$ as above. Then clearly $\mu_k > q^{1+\frac{1}{2}\epsilon}$ and moreover the first minimum λ_1 of the parallelepiped $|\mathbf{a}_j \mathbf{x}| \leqslant \mu_j$ $(1 \leqslant j \leqslant k)$ is at most $q^{-\epsilon/(2n)}$. But, on appealing again to the given inequality and applying the inductive hypothesis, we see that, if $q \gg 1$, then $\mu_j < 1$ for all $j < k$. Hence the proposition above shows that $\lambda_1 > \mu^{-\xi}$ for any $\xi > 0$ and any μ with $\mu \gg 1$ and $\mu \gg \mu_k$. Furthermore, by the case $n = 1$ of the theorem, we have $\mu_j > q^{-1}$ $(1 \leqslant j \leqslant n)$, whence $\mu_k \leqslant q^n$. Plainly the estimates for λ_1 are inconsistent if ξ is sufficiently small, and the contradiction proves the theorem.

Preliminary to the proof of the proposition, we observe that, with the notation of § 9, the linear forms $M_\tau = \mathbf{A}_\tau \mathbf{X}$ satisfy the hypotheses of § 7 with S given by those sets τ which include k. For it is easily verified from the Laplace expansions of $\mathbf{A}$ that, as σ runs through the complement of τ in $1, \dots, k$, the forms $\mathbf{A}_\sigma \mathbf{X}$ constitute the set adjoint to the M_τ, except possibly for a sign change; further, if σ does not include k, we have

$$\mathbf{A}_\sigma \mathbf{X} = X_\sigma + \Sigma(\pm \alpha_j) X_{\sigma-j+k},$$

where the summation is over all j in σ, on the right there occur the co-ordinates of $\mathbf{X}$, and $\sigma - j + k$ denotes the set σ with k in place of j. By hypothesis $1, \alpha_1, \dots, \alpha_n$ are linearly independent over the rationals, and thus we see that $\mathbf{A}_\sigma \mathbf{X} \neq 0$ for all integer vectors $\mathbf{X} \neq 0$, as required.

The proof of the proposition proceeds by induction on k; the result plainly holds for $k = 2$ by Lemma 5, and we assume now that it has been verified for all values up to $k-1$. Let l be any integer with $1 \leqslant l < k$ and, for any set τ of l distinct integers from $1, \dots, k$, let $\mu_\tau = \Pi \mu_j$, where the product is over all j in τ. By Lemma 7 we see that the successive minima $\nu_1, \dots, \nu_m$ of the parallelepiped $|M_\tau| \leqslant \mu_\tau$ satisfy $\nu_{m-1} \gg \nu_m \mu^{-k\zeta}$ for any $\zeta > 0$, provided that $\mu \gg 1$, $\mu \gg \mu_\tau$ for all τ and $\nu_1 \mu_\tau > \mu^{-\zeta}$ for τ in S. Further, with the notation of Lemma 8, it is clear that τ_m and τ_{m-1} consist of the integers $k-l+1, k-l+2, \dots, k$ and $k-l$, $k-l+2, \dots, k$ respectively. Thus, under the above conditions, we have

$$\lambda_{k-l}\lambda_{k-l+2} \dots \lambda_k \gg \lambda_{k-l+1} \dots \lambda_k \mu^{-k\zeta},$$

that is $\lambda_{k-l} \gg \lambda_{k-l+1}\mu^{-k\zeta}$. The required inequality $\lambda_1 > \mu^{-\xi}$ follows on applying the latter with $l = 1, 2, \dots, k-1$, noting that $\lambda_k \gg 1$, and taking ζ sufficiently small.

Since evidently $\mu_\tau \leqslant \mu_k$ for all τ, it remains only to prove that, for τ in S, $\nu_1 \mu_\tau > \mu^{-\zeta}$ for any μ with $\mu \gg 1$ and $\mu \gg \mu_k$. In fact it suffices to show that $\lambda_1 \mu_\tau^{1/l} > \mu^{-\zeta}$, for, again from Lemma 8, we have $\nu_1 \gg \lambda_1 \dots \lambda_l \geqslant \lambda_1^l$. Now, by the definition of λ_1, the parallelepiped $|\mathbf{a}_j \mathbf{x}| \leqslant \lambda_1 \mu_j$ $(1 \leqslant j \leqslant k)$

contains an integer point $\mathbf{x} \neq 0$; in fact the kth co-ordinate of $\mathbf{x}$ is not 0 since $\lambda_1 \leqslant 1$ by Minkowski's linear forms theorem, whence

$$\lambda_1 \mu_j < 1 \quad (1 \leqslant j < k),$$

and $\mathbf{a}_j\mathbf{x}$ is simply the jth co-ordinate of $\mathbf{x}$ when the kth co-ordinate vanishes. It follows that, if τ is any element of S, then the parallelepiped in R^l given by $|\mathbf{a}_i\mathbf{x}| \leqslant \lambda_1 \mu_i$, where i is restricted to τ and the co-ordinates of $\mathbf{x}$ with suffixes not in τ are disregarded, also contains an integer point $\mathbf{x} \neq 0$. Hence the first minimum λ_1' of the parallelepiped $|\mathbf{a}_i\mathbf{x}| \leqslant \mu_i'$ in R^l, where $\mu_i' = \mu_i/\mu_\tau^{1/l}$, is at most $\lambda_1 \mu_\tau^{1/l}$. It is therefore enough to prove that $\lambda_1' > \mu^{-\zeta}$; but this follows from the inductive hypothesis since clearly $\Pi \mu_i' = 1$ and $\mu_\tau > 1$. The theorem is herewith established.

8

MAHLER'S CLASSIFICATION

1. Introduction

A classification of the set of all transcendental numbers into three disjoint aggregates, termed S-, T- and U-numbers, was introduced by Mahler[†] in 1932, and it has proved to be of considerable value in the general development of the subject. The first classification of this kind was outlined by Maillet[‡] in 1906, and others were described by Perna[§] and Morduchai-Boltovskoj;[‖] but to Mahler's classification attaches by far the most interest.

As in the previous chapter, we define the height of a polynomial as the maximum of the absolute values of its coefficients, and we shall speak of the height only for polynomials with integer coefficients, not all 0. Let now ξ be any complex number, and for each pair of positive integers n, h, let $P(x)$ be a polynomial with degree at most n and height at most h for which $|P(\xi)|$ takes the smallest positive value; and define $\omega(n, h)$ by the equation $|P(\xi)| = h^{-n\omega(n,h)}$. Further define

$$\omega_n = \limsup_{h\to\infty} \omega(n, h), \quad \omega = \limsup_{n\to\infty} \omega_n,$$

and let ν be the least positive integer n for which $\omega_n = \infty$, writing $\nu = \infty$ if, in fact, $\omega_n < \infty$ for all n. Mahler characterizes the set of all complex numbers as follows:

$$\begin{array}{ll} A\text{-number} & \omega = 0, \ \nu = \infty, \\ S\text{-number} & 0 < \omega < \infty, \ \nu = \infty, \\ T\text{-number} & \omega = \infty, \ \nu = \infty, \\ U\text{-number} & \omega = \infty, \ \nu < \infty. \end{array}$$

We shall prove in §2 that the A-numbers are just the algebraic numbers; thus a transcendental number ξ is an S-number if $\omega(n, h)$ is uniformly bounded for all n, h, a U-number if, for some n, $\omega(n, h)$ is unbounded, and a T-number otherwise. Further we have:

† *J.M.* **166** (1932), 118–36.
‡ Bibliography.
§ *Giorn. Mat. Battaglini*, **52** (1914), 305–65.
‖ *Mat. Sbornik*, **41** (1934), 221–32.

Theorem 8.1. *Algebraically dependent numbers belong to the same class.*

Theorem 8.2. *Almost all numbers are S-numbers.*

Here 'almost all' is interpreted in the sense of Lebesgue measure theory, the linear and planar measures being taken for the real and complex numbers respectively.

The integer ν defined above is called the degree of ξ. It is clear that the Liouville numbers, mentioned in Chapter 1, are U-numbers of degree 1, and LeVeque† proved in 1953 the existence of U-numbers of each degree; we shall establish the latter in § 6. For many years it was an open question whether any T-numbers existed but, in 1968, an affirmative answer was obtained by Schmidt‡ on the basis of Wirsing's early version of Theorem 7.2, and this will be the theme of § 7. It is customary to subclassify the S-numbers according to 'type', defined as the supremum of the sequence $\omega_1, \omega_2, \ldots$. We shall show in § 2 that, for any transcendental ξ, ω_n is at least 1 or $\frac{1}{2}(1-1/n)$ according as ξ is real or complex, whence the type of ξ is respectively at least 1 or $\frac{1}{2}$. In 1965, Sprindžuk, confirming a conjecture of Mahler, proved that almost all real and complex numbers are S-numbers of type 1 and $\frac{1}{2}$ respectively. Moreover it was recently demonstrated by a refinement of this result that there exist S-numbers of arbitrarily large type. Thus, apart from a small gap in the kind of T-numbers that have so far been exhibited, the transcendental spectrum is, in a sense, complete. The latter measure-theoretical propositions will be the topic of the next chapter.

In the light of Theorem 8.2, one would expect any naturally defined number such as e, π, e^{π} and $\log \alpha$ for algebraic α not 0 or 1 to be an S-number. In 1929, Popken proved that indeed e is an S-number of type 1, and we shall confirm the result in Chapter 10. Theorem 3.1 shows that π, and in fact any non-vanishing linear combination of logarithms of algebraic numbers with algebraic coefficients, is either an S- or a T-number, but the latter possibility has not, as yet, been excluded. From the case $n = 1$ of Theorem 7.1 one sees, for instance, that $\sum_{n=1}^{\infty} a^{-b^n}$ is transcendental for any integers $a \geqslant 2$, $b \geqslant 3$, and, in the same context, Mahler§ proved in 1937 that also the decimal $\cdot 1234\ldots$,

† *J. London Math. Soc.* **28** (1953), 220–9.
‡ *Symposia Math. IV* (Academic Press, 1970), pp. 3–26.
§ *N.A.W.* **40** (1937), 421–8.

where the natural numbers are written in ascending order, is transcendental; and here again it has been proved that these are either S- or T-numbers.† For e^{π}, however, the possibility that it is a Liouville number has not even been excluded at present. Note that, by virtue of Theorem 8.1, the above results enable one to furnish many examples of algebraically independent numbers; indeed if ξ is any U-number, such as for instance $\Sigma 10^{-n!}$, and if η is, say, e or π or $\Sigma 10^{-10^n}$ or Mahler's decimal, then certainly ξ, η are algebraically independent.

In 1939, Koksma introduced a classification closely analogous to that of Mahler, which has also proved illuminating.‡ Let ξ be any complex number and for each pair of positive integers n, h, let α be an algebraic number with degree at most n and height at most h such that $|\xi-\alpha|$ takes the smallest positive value; and define $\omega^*(n,h)$ by the equation

$$|\xi-\alpha| = h^{-n\omega^*(n,h)-1}.$$

Koksma classified the complex numbers as A^*-, S^*-, T^*- or U^*-numbers in the same way as Mahler, but with ω^* in place of ω. Thus a transcendental number ξ is an S^*-number if $\omega^*(n,h)$ is uniformly bounded, a U^*-number if, for some n, $\omega^*(n,h)$ is unbounded, and a T^*-number otherwise. There is an exact correspondence between the two classifications, the S^*-, T^*- and U^*-classes being in fact identical with the S-, T- and U-classes respectively; moreover, the functions ω_n and ω_n^* take comparable values. Indeed it is easily verified that $\omega_n^* \leqslant \omega_n$, and simple lower bounds for ω_n^* in terms of ω_n were obtained by Wirsing.§ These imply, in particular, that $\omega_n^* = 1$ when $\omega_n = 1$, whence, in view of Sprindžuk's theorem, we have $\omega_n^* = 1$ for almost all real ξ. But it remains an open question whether $\omega_n^* \geqslant 1$ for all real ξ.

2. *A*-numbers

We prove here that the A-numbers are just the algebraic numbers. Suppose first that ξ is a real transcendental number. We consider the set of all numbers $Q(\xi)$, where Q denotes a polynomial, not identically 0, with degree at most n and with integer coefficients between 0 and h inclusive. The set evidently contains $(h+1)^{n+1}-1$ elements each with absolute value at most ch for some $c = c(n,\xi)$. If now we divide the interval $[-ch, ch]$ into h^{n+1} disjoint subintervals each of length $2ch^{-n}$, then there will be two distinct numbers $Q_1(\xi)$ and $Q_2(\xi)$ in the same

† *Acta Math.* **111** (1964), 97–120.
‡ For references and further discussion see Schneider (Bibliography).
§ *J.M.* **206** (1961), 67–77.

subinterval. Thus the polynomial $P = Q_1 - Q_2$ satisfies $|P(\xi)| < 2ch^{-n}$ and so $\omega_n \geq 1$. Similarly, if ξ is complex, we divide the intervals $[-ch, ch]$ on the real and imaginary axes into at most $h^{\frac{1}{2}(n+1)}$ disjoint subintervals each of length at most $c'h^{-\frac{1}{2}(n-1)}$ for some $c' = c'(n, \xi)$, and there will be two distinct numbers $Q_1(\xi)$ and $Q_2(\xi)$ with real and imaginary parts in the same subintervals. Thus we have

$$\omega_n \geq \tfrac{1}{2}(1 - 1/n).$$

Now if ξ is algebraic with degree m, then for any polynomial P as above, $P(\xi)$ is an algebraic number with degree at most m and height at most ch for some $c = c(n, \xi)$. Hence either $P(\xi) = 0$ or $|P(\xi)| > c'h^{-m}$ for some $c' = c'(n, \xi) > 0$. It follows that $n\omega(n, h)$ is uniformly bounded for all n, h, and this proves the assertion.

3. Algebraic dependence

Our purpose here is to prove Theorem 8.1. Suppose that ξ, η are algebraically dependent. Then they satisfy an equation $Q(\xi, \eta) = 0$, where $Q(x, y)$ is a polynomial with, say, degree k in x, l in y, and with algebraic coefficients, not all 0. Without loss of generality we can suppose that ξ, η are transcendental, for otherwise they would both be algebraic and so belong to the same class; also we can suppose that the coefficients of Q are rational integers, for this can evidently be ensured by taking, in place of Q, a product of its conjugates. Moreover we can suppose that all the zeros $\xi_1 = \xi, \xi_2, \ldots, \xi_k$ of $Q(x, \eta)$ are transcendental; for if one of these were algebraic then its minimal defining polynomial, say $p(x)$, would divide all the coefficients of $Q(x, y)$ regarded as a polynomial in y, and it would therefore suffice to consider $Q(x, y)/p(x)$ in place of $Q(x, y)$.

Let now P and $\omega(n, h)$ be defined as at the beginning of § 1 and put

$$J = P(\xi_1) \ldots P(\xi_k).$$

Clearly we have $$|J| \leq c_1 h^{-n\omega(n, h)+k-1},$$

where c_1, like c_2, c_3 below, depends only on ξ, η, n and Q. Further, J is symmetric in $\xi_1, \ldots, \xi_k$ and so, by the fundamental theorem on symmetric functions, it can be expressed as a polynomial in the elementary symmetric functions with total degree at most n and with height at most $c_2 h^k$. Now each elementary symmetric function is given by $\pm q_j/q_0$, where

$$Q(x, \eta) = q_0(\eta)\, x^k + q_1(\eta)\, x^{k-1} + \ldots + q_k(\eta).$$

Hence $q_0^n J$ is a polynomial in η with degree at most ln and height at most $h' = c_3 h^k$. If therefore $\omega'(n, h')$, ω'_n and ω' are defined for η in the same way as $\omega(n, h)$, ω_n and ω were defined for ξ, we have

$$h'^{ln\omega'(ln, h')} \geqslant c_1 h^{n\omega(n, h) - k + 1}.$$

This gives $kln\omega'_{ln} \geqslant n\omega_n - k + 1$, whence $kl\omega' \geqslant \omega$. Similarly, on interchanging ξ and η we obtain $kl\omega \geqslant \omega'$ and Theorem 8.1 follows.

4. Heights of polynomials

We establish now two lemmas which will be employed in the proof of Theorem 8.2 and in the next chapter. The propositions will be proved for polynomials with arbitrary complex coefficients, and here no restriction will attach to the definition of the height. $P(x)$ will denote a polynomial with degree n and height h, and constants implied by $\ll$ or $\gg$ will depend only on n.

Lemma 1. *For some integer j with $0 \leqslant j \leqslant n$ we have*

$$h \ll |P(j)| \ll h.$$

Proof. It is readily verified that

$$P(x) = \sum_{j=0}^{n} \frac{P(j)\, A(x)}{A'(j)\,(x-j)},$$

where $A(x) = x(x-1)\ldots(x-n)$, and A' denotes the derivative of A. Now we have $|A'(j)| \geqslant 1$, and clearly also the coefficients in the polynomials $A(x)/(x-j)$ are $\ll 1$. Thus we see that $|P(j)| \gg h$ for some j, and obviously we have $|P(j)| \ll h$ for all j. This proves the lemma.

Lemma 2. *If $P = P_1 P_2 \ldots P_k$, where P_i is a polynomial with height h_i, then*

$$h_1 h_2 \ldots h_k \ll h \ll h_1 h_2 \ldots h_k.$$

Proof. The right-hand estimate follows at once from the observation that every coefficient in P can be expressed as a sum of $\ll 1$ terms each given by a product of k coefficients, one from each of the P_i.

To establish the left-hand estimate, we begin by choosing an integer j to satisfy Lemma 1, and we denote by H_i the height of the polynomial $P_i(x+j)$. It is clear, on expressing $P_i(x)$ as a polynomial in $x-j$, that $h_i \ll H_i$. Now if η is any zero of $P(x+j)$, we deduce from the mean value theorem

$$h \ll |P(j)| = |P(\eta+j) - P(j)| = |\eta|\,|P'(\xi+j)|$$

for some ξ with $|\xi| \leqslant |\eta|$. Hence if $|\eta| < 1$, we have $h \ll |\eta| h$, that is $|\eta| \gg 1$. But the zeros of $P_i(x+j)$ are included in those of $P(x+j)$, and each coefficient in $P_i(x+j)$ can be written as the product of the constant coefficient $P_i(j)$ together with an elementary symmetric function in the reciprocals of the zeros. Thus we obtain $|P_i(j)| \gg H_i$, and the lemma follows since $P(j) = P_1(j) \dots P_k(j)$.

5. *S*-numbers

We proceed now to prove Theorem 8.2 for complex numbers in terms of planar Lebesgue measure; the argument for real numbers is similar. Again we shall speak of the height only for polynomials with integer coefficients.

We note first that if ξ is any complex number and P is any irreducible polynomial with degree at most n and height at most h, then the nearest zero α of P to ξ satisfies

$$|\xi-\alpha| \leqslant 2^n |P(\xi)|\, |P'(\alpha)|^{-1};$$

for if α' is any other zero of P we have

$$|\alpha-\alpha'| \leqslant |\xi-\alpha|+|\xi-\alpha'| \leqslant 2|\xi-\alpha'|.$$

Further we observe that $|P'(\alpha)| \gg h^{-n}$; for if p denotes the leading coefficient of P and if $\alpha_1, \dots, \alpha_m$ are any distinct conjugates of α then, on applying Lemma 2 with P_i given by $x-\alpha_i$, one sees that the algebraic integer† $p\alpha_1 \dots \alpha_m$ is $\ll h$, whence the norm of $P'(\alpha)$ multiplied by p^{n-1} is $\ll h^n |P'(\alpha)|$. If now ξ is a T- or U-number then, by Lemma 2, there exist, for some n, infinitely many polynomials P as above such that $|P(\xi)| < h^{-4n}$, and so the nearest zero α of P to ξ satisfies $|\xi-\alpha| \ll h^{-3n}$. Hence every T- and U-number belongs to the elements of infinitely many sets $S(n, h)$ for some n, where $S(n, h)$ consists of all discs centred on the algebraic numbers with degree at most n and height at most h, and with radius h^{-2n}. But there are $\ll h^{n+1}$ elements in each $S(n, h)$ and thus their total area is $\ll h^{-2}$. Since Σh^{-2} converges, it follows that the set of all T- and U-numbers has measure zero, as required.

6. *U*-numbers

We establish here the existence of U-numbers of each degree. In fact we shall show that, for any positive integer n, $\zeta^{1/n}$ is a U-number of degree n, where $\zeta = \frac{1}{3} + \sum_{m=1}^{\infty} 10^{-m!}$. Indeed we shall prove, more

† It is well known that this is an algebraic integer; see e.g. Hecke (Bibliography).

generally, that ξ is a U-number of degree n if there exists a sequence $\alpha_1, \alpha_2, \ldots$ of distinct algebraic numbers, with degree n, satisfying

$$|\xi - \alpha_j| < h_j^{-\omega_j}, \tag{1}$$

where h_j denotes the height of α_j and $\omega_j \to \infty$ as $j \to \infty$, provided that, for some $r \geqslant 1$, we have

$$h_j < h_{j+1} < h_j^{r\omega_j} \tag{2}$$

for all sufficiently large j. Clearly $\xi = \zeta^{1/n}$ satisfies (1) and (2) with $\alpha_j = (p_j/q_j)^{1/n}$, where

$$p_j = 10^{j!}\left(1 + 3\sum_{m=1}^{j} 10^{-m!}\right), \quad q_j = 3\,.\,10^{j!},$$

and with $\omega_j = j, r = 2$; also α_j has exact degree n since q_j is not a perfect power.

It suffices to show that if (1) and (2) hold then there are only finitely many algebraic numbers β with degree at most $n-1$ satisfying

$$|\xi - \beta| < b^{-(2n)^4 r}, \tag{3}$$

where b denotes the height of β. For then n is the least positive integer for which there exist sequences $\alpha_1, \alpha_2, \ldots$ and $\omega_1, \omega_2, \ldots$ as above satisfying (1), whence ξ is a U^*-number of degree n and so also a U-number of the same degree. To verify this connexion between U- and U^*-numbers, note that if $P_j(x)$ is the minimal defining polynomial of α_j then (1) gives, for all sufficiently large j,

$$|P_j(\xi)| \ll h_j^{-\omega_j+n} \leqslant h_j^{-\frac{1}{2}\omega_j},$$

where the implied constant depends only on ξ and n, and, conversely, if there were a sequence of polynomials $P_j(x)$ $(j = 1, 2, \ldots)$ with degree at most $n-1$ and height at most h_j such that $|P_j(\xi)| < h_j^{-\omega_j}$ then the nearest zero α_j of P_j to ξ would satisfy (1) with ω_j replaced by ω_j/n.

Now suppose that β is an algebraic number with degree at most $n-1$ such that (3) holds, and let j be the integer which, for b sufficiently large, satisfies

$$h_j < b^{4n^2 r} < h_{j+1}; \tag{4}$$

in the sequel we shall write briefly α, h, ω for α_j, h_j, ω_j. From (1) and (3) we have

$$|\alpha - \beta| \leqslant |\xi - \alpha| + |\xi - \beta| < h^{-\omega} + b^{-(2n)^4 r},$$

and, from (2) and (4), the terms on the right are at most $(bh)^{-2n^2}$, provided that $\omega > 4n^2$. On the other hand, $\alpha - \beta$ is a non-zero algebraic number with degree at most n^2, and each conjugate has absolute value $\ll bh$, where the implied constant depends only on n; further, the same estimate obtains for the leading coefficient in the minimal defining polynomial. Hence

$$|\alpha-\beta| \gg (bh)^{-n^2},$$

and thus we have a contradiction if b is sufficiently large; the contradiction establishes the result.

We remark finally that the inequality $|\alpha-\beta| \gg (ab)^{-n^2}$ implicit in the above argument, where α, β denote distinct algebraic numbers with degrees at most n and heights a, b respectively, and the implied constant depends only on n, can be much improved. Indeed, by considering the norm of $\alpha-\beta$ and using the result employed in § 5 on products of conjugates of algebraic numbers, one easily obtains $|\alpha-\beta| \gg a^{-l}b^{-m}$, where l, m denote the degrees of the fields generated by β over $Q(\alpha)$ and α over $Q(\beta)$ respectively. A special case of the latter inequality was discovered by A. Brauer† in 1929, but, curiously, the full result was recorded only relatively recently.‡

7. *T*-numbers

These exist, as we now show. To begin with, let $\alpha_1, \alpha_2, \ldots$ be any non-zero algebraic numbers and let $\nu_1, \nu_2, \ldots$ be any real numbers exceeding 1. We shall prove that there exists a sequence $\gamma_1, \gamma_2, \ldots$ of non-zero numbers with γ_j/α_j rational such that, if h_j denotes the height of γ_j, then $H_{j+1} > 2H_j$, where $H_j = h_j^{\nu_j}$, and furthermore, γ_{j+1} lies in the interval I_j consisting of all real x with

$$\tfrac{1}{4}H_j^{-1} < x-\gamma_j < \tfrac{1}{2}H_j^{-1};$$

in addition, we shall show that the sequence can be chosen so that, for some numbers $\lambda_1, \lambda_2, \ldots$ between 0 and 1 exclusive, we have

$$|\gamma_j-\beta| > B^{-1}$$

for all algebraic numbers β with degree $n \leqslant j$ distinct from $\gamma_1, \ldots, \gamma_j$, where $B = \lambda_n^{-1}b^{(3n)^4}$ and b denotes the height of β. Clearly then, $\gamma_1, \gamma_2, \ldots$ tends to a limit ξ which satisfies $|\xi-\beta| \geqslant B^{-1}$ for all algebraic numbers β distinct from $\gamma_1, \gamma_2, \ldots$, and also

$$\tfrac{1}{4}H_j^{-1} \leqslant \xi-\gamma_j \leqslant H_j^{-1}$$

for all j. We now take $\nu_j = (3n_j)^4$, where n_j denotes the degree of α_j,

† *J.M.* **160** (1929), 70–99.

‡ For references and further work in this context see *Michigan Math. J.* **8** (1961), 149–59 (R. Güting).

and we select $\alpha_1, \alpha_2, \ldots$ so that the equation $n_j = n$ has infinitely many solutions for each positive integer n. Then ξ is a T^*-number and hence, by observations similar to those recorded in § 6, also a T-number.

We shall in fact construct $\gamma_1, \gamma_2, \ldots$ so that four further conditions are satisfied. Let J_j be the set of all x in I_j such that $|x-\beta| > 2B^{-1}$ for all algebraic numbers β with degree $n \leqslant j$ which are distinct from $\gamma_1, \ldots, \gamma_j$ and satisfy $B > H_j$. Then we shall ensure that (i) γ_j is in J_{j-1}, (ii) the measures of I_j and J_j satisfy $|J_j| > \frac{1}{2}|I_j|$, (iii) we have $|\gamma_j - \beta| > 2B^{-1}$ for all $\beta \neq \gamma_j$ with degree j, (iv) if $\gamma_j/\alpha_j = p_j/q_j$ as a fraction in its lowest terms, with $q_j > 0$, then $|\gamma_j - \beta| > q_j^{-1}$ for all β with degree $n \leqslant j$ and with $b^{3n} \leqslant q_j$.

To define γ_1, we note first that, for every large prime q_1, there are $\gg q_1$ numbers γ of the form $(p_1/q_1)\alpha_1$ in the interval $(1, 2)$, where the implied constant depends only on α_1, and these have mutual distances $\gg q_1^{-1}$. Further, there are $\ll q_1^{\frac{2}{3}}$ rationals β with $b^3 \leqslant q_1$ and so there are $\ll q_1^{\frac{2}{3}}$ numbers γ satisfying $|\gamma - \beta| \leqslant q_1^{-1}$ for at least one such β. We can therefore select γ_1 so that (iv) holds, and then, by Theorem 7.2, we can choose λ_1 so that the conditions concerning $|\gamma_1 - \beta|$ are satisfied. We shall show in a moment that also (ii) holds in the case $j = 1$ if $q_1 \gg 1$.

Now suppose that $\gamma_1, \ldots, \gamma_{j-1}$ have already been defined to satisfy the above conditions; we proceed to construct γ_j. Constants implied by $\ll$ or $\gg$ will depend only on the numbers so far specified, including possibly $\lambda_1, \ldots, \lambda_{j-1}$. First let J'_{j-1} be defined like J_{j-1} but with the additional restriction that the heights of the β in question satisfy $b^{3n} \leqslant q_j$. Clearly the number of β for whieh the latter inequality holds is $\ll q_j^{\frac{2}{3}}$ and so J'_{j-1} consists of $\ll q_j^{\frac{2}{3}}$ intervals. Further, J'_{j-1} includes J_{j-1} and so, by (ii), we have $|J'_{j-1}| \geqslant \frac{1}{2}|I_{j-1}| \gg 1$. It follows that, for any large prime q_j, there are $\gg q_j$ numbers γ in J'_{j-1} of the form $(p_j/q_j)\alpha_j$, where p_j is an integer $\ll q_j$ with $(p_j, q_j) = 1$. Furthermore, any such γ is in fact in J_{j-1}, for if the height of β satisfies $b^{3n} > q_j$ then $B > q_j^{(3n)^3}$ and thus, on noting that $(q_j/p_j)\beta$ has height $\ll q_j^n b$, we obtain from Theorem 7.2

$$|\gamma - \beta| \gg q_j^{-1}(q_j^n b)^{-3n} > 2B^{-1}.$$

Now, as above, there are $\ll q_j^{\frac{2}{3}}$ numbers β satisfying the hypotheses of (iv) and hence one can select $\gamma = \gamma_j$ in J_{j-1} so that this condition is valid. Then clearly we have $|\gamma_j - \beta| > B^{-1}$ for all β distinct from $\gamma_1, \ldots, \gamma_{j-1}$ with degree $n < j$ and with $B > H_{j-1}$; and indeed this holds also for $B \leqslant H_{j-1}$, for then, taking k as the least suffix $\geqslant n$ for which $B \leqslant H_k$ and appealing to (i) or (iii) with $j = k$ according as $k > n$ or

$k = n$, we obtain

$$|\gamma_j - \beta| \geqslant |\gamma_k - \beta| - |\gamma_j - \gamma_k| > 2B^{-1} - H_k^{-1} \geqslant B^{-1}.$$

We now use Theorem 7.2 and choose λ_j so that $|\gamma_j - \beta| > 2B^{-1}$ for all algebraic numbers $\beta \neq \gamma_j$ with degree $n = j$.

It remains only to show, as in the case $j = 1$, that (ii) will be satisfied if q_j is sufficiently large. Now we have $|x - \beta| > 2B^{-1}$ for all x in I_j and all $\beta \neq \gamma_j$ with degree $n \leqslant j$ and with $H_j < B \leqslant H_j^3$. For if $b^{3n} \leqslant q_j$ then, since $H_j \gg q_j^{\nu_j}$ and $\nu_j > 1$, it follows from (iv) that

$$|x - \beta| \geqslant |\gamma_j - \beta| - |\gamma_j - x| \geqslant q_j^{-1} - H_j^{-1} \geqslant 2H_j^{-1} > 2B^{-1},$$

and if $b^{3n} > q_j$ then, on appealing again to Theorem 7.2, we obtain

$$|x - \beta| \geqslant q_j^{-4n^3} b^{-3n} - H_j^{-1} \geqslant B^{-\frac{1}{4}} - B^{-\frac{1}{3}} > 2B^{-1}.$$

Hence any x in the complement of J_j in I_j satisfies $|x - \beta| \leqslant 2B^{-1}$ for some β with degree $n \leqslant j$ and with $B > H_j^3$. But the number of β with degree n and height b is $\ll b^n$, and so the complement has measure $\ll \Sigma B^{-1} b^n$, where the sum is over all n, b with $n \leqslant j$ and $B > H_j^3$. This is plainly $\ll H_j^{-2} < \frac{1}{8} H_j^{-1}$, and the required result follows.

It will be seen that the above argument allows one to construct a T-number with ω_n taking any value $\geqslant (3n)^4$. This can easily be reduced to a bound of order n^2, but at present, apparently, not readily to one of order n as would be needed to fill the spectrum.

9

METRICAL THEORY

1. Introduction

As remarked previously, Mahler conjectured in 1932 that almost all real numbers are S-numbers of type 1 and almost all complex numbers are S-numbers of type $\frac{1}{2}$.† He originally proved that, certainly, they are both of type at most 4, and 4 was reduced to 3 and $\frac{5}{2}$ in the real and complex cases respectively by Koksma in 1939. LeVeque improved these in 1953 to 2 and $\frac{5}{3}$, and Volkmann further reduced them in 1964 to $\frac{4}{3}$ and $\frac{2}{3}$. Moreover, proofs of Mahler's conjecture in the special cases with $n = 2$ and $n = 3$ were given by Kubilyus, Kasch and Volkmann. Finally, in 1965, Sprindžuk‡ obtained a complete proof of Mahler's conjecture for all n, and indeed with the best possible value of ω_n.

We shall establish here a refinement of Sprindžuk's result which was derived by the author in 1966.§ Denoting by $\psi(h)$ a positive monotonic decreasing function of the integer variable $h > 0$ such that $\Sigma\psi(h)$ converges, we prove:

Theorem 9.1. *For almost all real θ and any positive integer n there exist only finitely many polynomials P with degree n and integer coefficients such that $|P(\theta)| < (\psi(h))^n$, where h denotes the height of P.*

A similar result holds for almost all complex numbers θ with the exponent n replaced by $\frac{1}{2}(n-1)$. It is clear from, for instance, Minkowski's linear forms theorem, that the assertion would not remain valid with $\psi(h) = 1/h$, and indeed it is easily verified that almost no θ would have the properties required in the case $n = 1$ if $\Sigma\psi(h)$ were divergent. But it seems likely that the function $(\psi(h))^n$ can be replaced by $h^{-n+1}\psi(h)$, and this conjecture has in fact been established for $n \leqslant 3$.

The theorem has recently been applied to evaluate the Hausdorff dimension of certain sets; in particular, it has been employed to show that, for any $\lambda \geqslant 1$ and any positive integer n, the set of all real ξ such that, for any $\lambda' < \lambda$, there exist infinitely many algebraic numbers β

† *M.A.* **106** (1932), 131–9.
‡ Bibliography; this contains references to the earlier works.
§ *Proc. Roy. Soc. London*, A **292** (1966), 92–104.

with degree at most n satisfying $|\xi-\beta| < b^{-(n+1)\lambda'}$, where b denotes the height of β, has dimension $1/\lambda$. This generalizes a well-known theorem of Jarník and Besicovitch; and it immediately implies the result mentioned in the last chapter on the existence of S-numbers of arbitrarily large type.†

Various avenues for further investigation are suggested by the work here. For instance it would be of interest to obtain results analogous to Theorem 9.1 for polynomials in several variables, and in fact some progress in this connexion, more especially for cubic polynomials in two unknowns, has been made by R. Slesoraitene.‡ In another direction, it follows from Theorem 9.1, by a classical transference principle, that, for any $\epsilon > 0$ and any positive integer n, there exist, for almost all real θ, only finitely many positive integers q such that

$$\max \|q\theta^j\| < q^{-(1/n)-\epsilon} \quad (1 \leqslant j \leqslant n),$$

and this raises the problem of confirming the stronger proposition in which the above inequality is replaced by

$$q^{1+\epsilon}\|q\theta\| \ldots \|q\theta^n\| < 1,$$

where the notation is that of Theorem 7.1. The problem seems quite difficult.

2. Zeros of polynomials

We record here, for later reference, some simple inequalities concerning the distances between the zeros of polynomials. Let $P(x)$ be a polynomial with degree n and distinct zeros $\kappa_1, \ldots, \kappa_n$. We note first that if θ is any real number with $|\theta-\kappa_1| \leqslant |\theta-\kappa_j|$ for all j then

$$|P(\theta)| \geqslant 2^{-n}|P'(\kappa_1)|\,|\theta-\kappa_1|, \tag{1}$$

where P' denotes the derivative of P. For clearly $|\kappa_1-\kappa_j| \leqslant 2|\theta-\kappa_j|$, and we have

$$P'(\kappa_1) = a(\kappa_1-\kappa_2)\ldots(\kappa_1-\kappa_n),$$

where a denotes the leading coefficient of P. Similarly we obtain

$$|P(\theta)|\,|\kappa_1-\kappa_2| \geqslant 2^{-n}|P'(\kappa_1)|\,|\theta-\kappa_1|^2. \tag{2}$$

Further we observe that if $|\theta-\kappa_1| \leqslant |\kappa_1-\kappa_j|$ for all $j \geqslant 2$ then $|\theta-\kappa_j| \leqslant 2|\kappa_1-\kappa_j|$ and so

$$|P(\theta)| \leqslant 2^n|P'(\kappa_1)|\,|\theta-\kappa_1|. \tag{3}$$

† *Proc. London Math. Soc.* **21** (1970), 1–11 (A. Baker and W. M. Schmidt).

‡ See various papers in *Litovsk. Mat. Sb.* since 1969; see also Sprindžuk's address in *Actes, Congrès international math.* (1970).

Now suppose that $P(x)$, $Q(x)$ are polynomials with integer coefficients and degree $n \geqslant 2$; let their leading coefficients be a, b and their zeros be $\kappa_1, \ldots, \kappa_n$ and $\lambda_1, \ldots, \lambda_n$ respectively, all of which are supposed to be distinct and have absolute values at most K. We shall write, for brevity,

$$p = |a^n(\kappa_1 - \kappa_3) \ldots (\kappa_1 - \kappa_n)|,$$

and we shall denote by q the analogous function of Q. Our purpose is to prove that if $|\kappa_1 - \kappa_2| \leqslant |\kappa_1 - \kappa_j|$ for all $j \geqslant 2$, if $|\kappa_1 - \kappa_2| < p^{-\frac{1}{2}}$, and if also the analogous inequalities hold for Q, then

$$|\kappa_1 - \lambda_1| \gg \min(p^{-\frac{1}{2}}, q^{-\frac{1}{2}}), \tag{4}$$

where the implied constant depends only on n and K.

For the proof, we suppose that (4) does not hold and we shall obtain a contradiction if the implied constant is sufficiently large. First we observe that $|\kappa_1 - \kappa_j| \gg p^{-\frac{1}{2}}$ for all $j \geqslant 3$. This is a consequence of the fact that the discriminant of P, namely

$$a^{2n-2} \prod_{i<j} (\kappa_i - \kappa_j)^2,$$

has absolute value at least 1; for, in view of the inequality $|\kappa_i - \kappa_j| \ll 1$ valid for all i, j, it follows that

$$|(\kappa_1 - \kappa_2)(\kappa_2 - \kappa_j)| \gg p^{-1} \quad (j \geqslant 3),$$

and, by hypothesis, we have

$$|\kappa_2 - \kappa_j| \leqslant 2|\kappa_1 - \kappa_j| \quad \text{and} \quad |\kappa_1 - \kappa_2| < p^{-\frac{1}{2}}.$$

Hence, from the converse of (4), we obtain $|\kappa_1 - \kappa_j| \gg |\kappa_1 - \lambda_1|$ and so

$$|\kappa_j - \lambda_1| \leqslant |\kappa_j - \kappa_1| + |\kappa_1 - \lambda_1| \ll |\kappa_j - \kappa_1|$$

for all $j \geqslant 3$. This gives

$$|a^{n-1} P(\lambda_1)| \ll p\,|(\kappa_1 - \lambda_1)(\kappa_2 - \lambda_1)|, \tag{5}$$

and, plainly, an analogous inequality holds for Q.

We now use the fact that the absolute value of the resultant of P and Q, namely $|ab|^n \prod |\kappa_i - \lambda_j|$, is at least 1. Since $|\kappa_i - \lambda_j| \ll 1$ this gives

$$|ab|^{n-1} |P(\lambda_1) Q(\kappa_1)(\kappa_2 - \lambda_2)(\kappa_1 - \lambda_1)^{-1}| \gg 1$$

and so, from (5) and its analogue for Q, we obtain

$$|(\kappa_1 - \lambda_1)(\kappa_1 - \lambda_2)(\kappa_2 - \lambda_1)(\kappa_2 - \lambda_2)| \gg (pq)^{-1}. \tag{6}$$

Further, by the converse of (4) and the hypothesis $|\kappa_1 - \kappa_2| \leqslant p^{-\frac{1}{2}}$ we have $|\kappa_2-\lambda_1| \ll p^{-\frac{1}{2}}$ and similarly $|\kappa_1-\lambda_2| \ll q^{-\frac{1}{2}}$. Furthermore we see that

$$|\kappa_2-\lambda_2| \leqslant |\kappa_2-\kappa_1|+|\kappa_1-\lambda_2| \ll \max(p^{-\frac{1}{2}}, q^{-\frac{1}{2}}).$$

But this together with (6) implies the validity of (4), contrary to supposition. The contradiction proves the assertion.

3. Null sets

Let now ψ be any function as in § 1 and, for any positive integer n and any real θ, let $\mathscr{P}(n, \psi, \theta)$ be the set of all polynomials P satisfying the hypotheses of Theorem 9.1. The theorem asserts that the set $\mathscr{R}(n, \psi)$ of all θ for which $\mathscr{P}(n, \psi, \theta)$ contains infinitely many elements has measure zero. We shall show here that it suffices to establish the following modified result.

The set $\mathscr{S}(n, \psi)$ of all θ for which $\mathscr{P}(n, \psi, \theta)$ contains infinitely many polynomials P that are (i) *irreducible and* (ii) *have leading coefficients which exceed the absolute values of the remaining coefficients, has measure zero.*

We begin by observing that, for any θ in $\mathscr{R}(n, \psi)$, there exists, by Lemma 1 of Chapter 8, an integer j with $0 \leqslant j \leqslant n$ such that infinitely many polynomials P in $\mathscr{P}(n, \psi, \theta)$ satisfy $|P(j)| \gg h$; and by taking $-P$ in place of P if necessary we can suppose that $P(j) > 0$. It clearly suffices to show that the set of θ in $\mathscr{R}(n, \psi)$ which corresponds to a fixed integer j has measure zero, and this is equivalent to proving that the translate, consisting of all numbers $\xi = \theta - j$, has measure zero. Now ξ satisfies $|P(\xi+j)| < (\psi(h))^n$ for all P in $\mathscr{P}(n, \psi, \theta)$, and $P(x+j)$ is a polynomial in x with height at most Ch for some C depending only on n. Further, there is a positive monotonic decreasing function $\sigma(h)$ such that $\Sigma\sigma(h)$ converges, $\sigma(h) \geqslant \psi(h)$ and $\sigma(h)/\sigma(Ch) \leqslant 2C^2$; indeed one can take $\sigma(1) = 2\psi(1)$ and

$$h(h-1)\,\sigma(h) = \sum_{k=2}^{h} (2k-2)\,\psi(k) \quad (h \geqslant 2),$$

whence

$$\sum_{h=1}^{n} \sigma(h) = 2n^{-1} \sum_{m=1}^{n} \sum_{h=1}^{m} \psi(h),$$

and so $\Sigma\sigma(h) = 2\Sigma\psi(h)$. Hence ξ is an element of $\mathscr{R}(n, \phi)$, where $\phi = 2C^2\sigma$, and infinitely many polynomials P in $\mathscr{P}(n, \phi, \xi)$ have constant coefficients exceeding ch for some $c > 0$, depending only on n. For any such P, the polynomial $Q(x) = x^n P(1/x)$ has leading coefficient exceeding ch and hence $R(x) = Q(c^{-1}x)$ satisfies (ii), assuming that

$c < 1$. Moreover, $R(x)$ has height at most $c^{-n}h$, and also integer coefficients if c^{-1} is an integer. Furthermore, for any positive integer k and any ξ as above with $|\xi| > k^{-1}$, the number $\eta = c\xi^{-1}$ satisfies

$$|R(\eta)| < (k\phi(h))^n.$$

It is plainly enough to prove that the set of all such η has measure zero; for given a covering of the η's by intervals $I_1, I_2, \ldots$, we obtain a covering $I'_1, I'_2, \ldots$ of the ξ's, where I'_j consists of all cx^{-1} with x in I_j and with $|x| > k^{-1}$, and clearly we have $|I'_j| \leqslant k^2|I_j|$. Thus, on utilizing again the above construction of σ, we see that it is necessary now only to show that the set $\mathscr{T}(n, \psi)$ of all θ for which $\mathscr{P}(n, \psi, \theta)$ contains infinitely many polynomials which satisfy (ii) but not necessarily (i), has measure zero.

Here we use induction. Clearly the sets $\mathscr{S}(1, \psi)$ and $\mathscr{T}(1, \psi)$ are identical and so the required result holds for $n = 1$. We assume that, for any ψ, the sets $\mathscr{R}(m, \psi)$ with $m < n$ are null and that also $\mathscr{S}(n, \psi)$ is null, and we proceed to prove that then each $\mathscr{T}(n, \psi)$ is null. For every θ in $\mathscr{T}(n, \psi)$, infinitely many P in $\mathscr{P}(n, \psi, \theta)$ satisfy (ii), and if infinitely many of these were irreducible then θ would be in $\mathscr{S}(n, \psi)$ and the required result would follow. Hence we shall suppose that all the P are reducible. Then each contains as a factor at least one polynomial Q with integer coefficients and degree $m < n$ satisfying $|Q(\theta)| < (\psi(h))^n$; further, infinitely many of the P correspond to a fixed integer m and, unless θ is algebraic, there will be infinitely many distinct polynomials among the associated Q. Now appealing to Lemma 2 of Chapter 8 and employing for a third time an averaging construction as above, we conclude that a function ϕ exists such that every θ in $\mathscr{T}(n, \psi)$ is in one at least of the sets $\mathscr{R}(m, \phi)$ with $m < n$. Each of these is null by the inductive hypothesis and so $\mathscr{T}(n, \psi)$ is null, as required.

4. Intersections of intervals

We establish here a further simple measure-theoretical result needed for the proof of Theorem 9.1.

For each positive integer h, let $\mathscr{U}(h)$ be a finite set of real closed intervals, and let $\mathscr{V}(h)$ be a subset of $\mathscr{U}(h)$ such that for each I in $\mathscr{V}(h)$ there is a $J \neq I$ in $\mathscr{U}(h)$ with $|I \cap J| \geqslant \frac{1}{2}|I|$. Further let W and w be the set of points contained in infinitely many $V(h)$ and in infinitely many $v(h)$ respectively, where $V(h)$ is the union of the points of the intervals

I of $\mathscr{V}(h)$, and $v(h)$ is that of the intervals $I \cap J$ with I in $\mathscr{V}(h)$ and $J \neq I$ in $\mathscr{U}(h)$. Our purpose is to prove that if w is null then so also is W. We have

$$w = \bigcap_{1 \leqslant m < \infty} \bigcup_{h \geqslant m} v(h),$$

and thus, if w is null, then, for any $\epsilon > 0$, there is an integer m such that, for all $n \geqslant m$, the union of the $v(h)$, taken over all h with $m \leqslant h \leqslant n$, has measure at most ϵ. Now this union consists of a finite set of disjoint intervals and, by the definition of $\mathscr{V}$, we see that the set obtained on expanding each of these intervals symmetrically about its centre to three times its length will cover all the $V(h)$ taken over the same range of h. Thus, for every $n \geqslant m$, the latter set has measure at most 3ϵ, and, on noting that W can be expanded like w above with V in place of v, the assertion follows.

5. Proof of main theorem

By virtue of § 3, it suffices to show that every set $\mathscr{S}(n, \psi)$ has measure zero. It is easily verified that $\mathscr{S}(1, \psi)$ is null and we shall assume that $\mathscr{S}(m, \psi)$ is null for $m < n$; we proceed to establish the result for $m = n \geqslant 2$.

Let $\mathscr{Q}(n, h)$ be the set of all polynomials with degree n, integer coefficients and height h satisfying (i) and (ii) of § 3. Further let $\kappa_1, \ldots, \kappa_n$ be the zeros of any element P of $\mathscr{Q}(n, h)$, and let

$$\tau_j = \min |\kappa_i - \kappa_j|,$$

where the minimum is taken over all $i \neq j$. By (i) we have $\tau_j > 0$ and from (ii) we obtain $|\kappa_j| \leqslant n$, since clearly

$$|P(x) - hx^n| \leqslant nh \max(1, |x|^{n-1}).$$

Suppose now that ψ is any function as in § 1, let

$$\nu_j = 2^n |P'(\kappa_j)|^{-1} (\psi(h))^n \quad (1 \leqslant j \leqslant n),$$

and let $I_j = I_j(P)$ be the interval (possibly empty) formed by the intersection of the real axis with the closed disc in the complex plane with centre κ_j and radius

$$\mu_j = \min\{\nu_j, (\tau_j \nu_j)^{\frac{1}{2}}\}.$$

From (1) and (2) we see that every element of $\mathscr{S}(n, \psi)$ is contained in infinitely many $\mathscr{I}_j(h)$ for some j, where $\mathscr{I}_j(h)$ denotes the set of all $I_j(P)$ as P runs through the elements of $\mathscr{Q}(n, h)$. We proceed to prove that the set of points contained in infinitely many $\mathscr{J}_1(h)$ has measure zero; the proof when $j > 1$ is similar and this will therefore suffice to establish the theorem. There is now no loss of generality in assuming that the zeros of P are so ordered that $\tau_1 = |\kappa_1 - \kappa_2|$.

We divide the polynomials P in $\mathscr{Q}(n,h)$ into two disjoint classes, placing P in $\mathscr{A}(n,h)$ if $\tau_1 \geqslant p^{-\frac{1}{2}}$ and in $\mathscr{B}(n,h)$ otherwise, where p is defined as in § 2, with $a = h$. We denote by $\mathscr{K}(h)$ and $\mathscr{L}(h)$ the union of all $I_1(P)$ as P runs through the elements of $\mathscr{A}(n,h)$ and $\mathscr{B}(n,h)$ respectively. Then clearly the union of $\mathscr{K}(h)$ and $\mathscr{L}(h)$ is just $\mathscr{J}_1(h)$ and it suffices to prove that the set $\mathscr{K}$ of points contained in infinitely many $\mathscr{K}(h)$ and likewise the set $\mathscr{L}$ of points contained in infinitely many $\mathscr{L}(h)$ have measure zero.

We prove first that $\mathscr{K}$ is null. Since $\psi(h)$ is positive monotonic decreasing and $\Sigma\psi(h)$ converges, we have $h\psi(h) \to 0$ as $h \to \infty$ and so there is no loss of generality in assuming that $\psi(h) < h^{-1}$ for all h. For each P in $\mathscr{A}(n,h)$, let $I = I(P)$ be the interval formed by the intersection of the real axis with the closed disc in the complex plane with centre κ_1 and radius $(\psi(h))^{-1}\mu_1$. Clearly $I_1 \subset I$ and $|I_1| \leqslant \psi(h)\,|I|$. We denote by $\mathscr{U}(h)$ the set of all $I(P)$ and by $\mathscr{V}(h)$ the maximal subset of $\mathscr{U}(h)$ possessing the property specified in § 4. Retaining the notation of that section, we proceed now to show that w is null. First we observe that every θ in $I(P)$ satisfies

$$|\theta - \kappa_1| \leqslant (\psi(h))^{-1}\nu_1 \leqslant h^{-n+1}|P'(\kappa_1)|^{-1} = |(\kappa_1 - \kappa_2)p|^{-1}, \qquad (7)$$

provided that h is sufficiently large; and the number on the right is at most $|\kappa_1 - \kappa_2|$ by the definition of $\mathscr{A}(n,h)$. Hence (3) holds and so

$$|P(\theta)| \leqslant 2^n|P'(\kappa_1)|\,(\psi(h))^{-1}\nu_1 = 2^{2n}(\psi(h))^{n-1}.$$

Now if θ were also a point of $I(Q)$ for some $Q \neq P$ in $\mathscr{A}(n,h)$ then the polynomial $R = P - Q$ would satisfy $|R(\theta)| \leqslant 2^{2n+1}(\psi(h))^{n-1}$. Further, from (ii), we see that R has degree at most $n-1$ and height at most $2h$. But, for every θ in w, there exist infinitely many distinct R with these properties and thus, on appealing again to the construction in § 3, it follows that w is contained in the union of sets $\mathscr{R}(m,\phi)$ for a suitable function ϕ, where $1 \leqslant m < n$. Our inductive hypothesis together with the result of § 3 shows that $\mathscr{R}(m,\phi)$ is null for each m, and hence w is null, as required.

We conclude from § 4 that W is null and thus to complete the proof that $\mathscr{K}$ is null it is necessary only to verify that the set of points in infinitely many $\mathscr{K}(h)$, with those $I_1(P)$ excluded for which the corresponding I is in $\mathscr{V}(h)$, has measure zero. Now if $I(P)$ and $I(Q)$ are distinct elements not contained in $\mathscr{V}(h)$ then

$$|I(P) \cap I(Q)| < \tfrac{1}{2} \min(|I(P)|, |I(Q)|).$$

This implies, as one readily verifies, that no point can be contained in three distinct intervals $I(P)$ not in $\mathscr{V}(h)$. Further, all $I(P)$ are included in $[-3n, 3n]$, for we have $|\kappa_j| \leqslant n$ and, as above, $|\theta-\kappa_1| \leqslant \tau_1$ for every θ in $I(P)$. Hence the total length of all $I(P)$ not in $\mathscr{V}(h)$ is at most $12n$. The corresponding $I_1(P)$ have therefore total length at most $12n\psi(h)$, and that $\mathscr{K}$ is null follows immediately since $\Sigma\psi(h)$ converges.

It remains to prove that $\mathscr{L}$ is null. For each positive integer k, let $\mathscr{C}(n,k)$ be the union of the sets $\mathscr{B}(n,h)$ with $4^{k-1} \leqslant h < 4^k$, and, for each integer l, let $\mathscr{C}(n,k,l)$ be the subset of $\mathscr{C}(n,k)$ consisting of all polynomials P with $4^{l-1} \leqslant p < 4^l$. Then, by (7), for each P in $\mathscr{C}(n,k,l)$, $I_1(P)$ has length at most

$$2\mu_1 \leqslant 2(\tau_1 \nu_1)^{\frac{1}{2}} \ll (4^{-l+1}\psi(4^{k-1}))^{\frac{1}{2}} \ll 2^{-l-k},$$

where the implied constants depend only on n. Further, if $I_1(P)$ is not empty then the imaginary part of κ_1 is at most μ_1. It follows from (4), on applying a simple box argument to the interval $[-n, n]$, that, if $k \gg 1$, then the number of polynomials P in $\mathscr{C}(n,k,l)$ for which $I_1(P)$ is not empty is $\ll 2^l+1$. Hence the total length of all $I_1(P)$ with P in $\mathscr{C}(n,k,l)$ is $\ll 2^{-k}(2^{-l}+1)$. But from the estimates in § 2 relating to the discriminant of P we see that $p \gg 1$, and clearly also $p \ll 4^{nk}$. Thus, for any n and k, the number of non-empty sets $\mathscr{C}(n,k,l)$ is $\ll k$, and, for such sets, we have $2^{-l} \ll 1$. We conclude that the total length of all $I_1(P)$ with P in $\mathscr{C}(n,k)$ is $\ll k2^{-k}$, and that $\mathscr{L}$ is null follows from the convergence of $\Sigma k2^{-k}$. This completes the proof of the theorem.

10

THE EXPONENTIAL FUNCTION

1. Introduction

In a classic memoir of 1899, Borel† obtained a refinement of Hermite's theorem on the exponential function and thereby established the first measure of transcendence for e. He proved that, for any positive integer n, there are only finitely many polynomials P with integer coefficients and degree n satisfying $|P(e)| < h^{-\phi(h)}$, where h denotes the height of P and $\phi(h) = c \log\log h$ for some $c = c(n) > 0$. Borel's result was much improved by Popken‡ in 1929; Popken showed that $\phi(h)$ can be replaced by $n + \epsilon(h)$, where $\epsilon(h) = c/\log\log h$ with $c = c(n) > 0$, and this plainly implies that e is an S-number of type 1. Mahler§ later derived an explicit expression for c of the form $c'n^2 \log(n+1)$, where now c' is absolute.

In 1965, a generalization of Popken's result similar to Theorem 7.1 was established by the author,|| and this will be the subject of the present chapter.

Theorem 10.1. *For any distinct, non-zero rationals $\theta_1, \ldots, \theta_n$ and any $\epsilon > 0$ there are only finitely many positive integers q such that*

$$q^{1+\epsilon}\|qe^{\theta_1}\| \ldots \|qe^{\theta_n}\| < 1.$$

The theorem plainly yields all the corollaries recorded after Theorem 7.1 with $\alpha_1, \ldots, \alpha_n$ replaced by $e^{\theta_1}, \ldots, e^{\theta_n}$, and indeed Theorem 7.2 holds with α replaced by e^{θ} for any non-zero rational θ. Furthermore, in contrast to the work of Chapter 7, the arguments here enable one to replace ϵ by a function $\epsilon(q)$ tending to 0 as $q \to \infty$, namely $c(\log\log q)^{-\frac{1}{2}}$ where $c = c(\theta_1, \ldots, \theta_n) > 0$.

The proof of the theorem involves techniques similar to those introduced by Siegel in his studies on the Bessel functions, which will be discussed in the next chapter. In particular, Dirichlet's box principle will be employed to construct certain linear forms in $e^{\theta_1 x}, \ldots, e^{\theta_n x}$ with polynomial coefficients that vanish to a high order at the origin. Linear forms of this kind occurred in the works of Popken and Mahler, but

† *C.R.* **128** (1899), 596–9. ‡ *M.Z.* **29** (1929), 525–41.
§ *J.M.* **166** (1932), 118–50. || *Canadian J. Math* **17** (1965), 616–26.

there they were derived explicitly by means of analytic integrals. Clearly Theorem 10.1 improves upon the Popken–Mahler theorem except when the polynomial P has coefficients that are, in absolute value, nearly all equal, and then the earlier work is slightly stronger in view of the more rapidly decreasing function ϵ. Feldman† has shown that the techniques used here furnish a function $\epsilon(q)$ of order $(\log\log q)^{-1}$ for certain series closely related to the exponential function.

The arguments of this chapter do not extend easily to furnish Theorem 10.1 for algebraic $\theta_1, \ldots, \theta_n$. Some results in this context were obtained in the original paper of Mahler, but they would seem to be far from best possible. In fact, even in the most precise analogue of Theorem 7.2 established to date, taking $\alpha = e^\theta$ with θ algebraic, the exponent of B tends rapidly to $-\infty$ as the degree of θ increases.‡ Nevertheless, a construction similar to that employed in § 2 below yields at once a negative answer to the power series analogue of a well-known problem of Littlewood. Littlewood asked whether, for any real θ, ϕ and any $\epsilon > 0$, there exists a positive integer q such that

$$q\,\|q\theta\|\,\|q\phi\| < \epsilon;$$

the series $\theta = e^{1/x}$, $\phi = e^{2/x}$ provide a counter-example to the analogue,§ but the problem itself remains unsolved. And the latter recalls to mind another outstanding question in Diophantine approximation, namely whether every continued fraction with unbounded partial quotients is necessarily transcendental; this too seems very difficult.

2. Fundamental polynomials

We suppose that $\theta_1, \ldots, \theta_n$ are distinct rationals and that $0 < \epsilon < 1$. Constants implied by $\ll$ or $\gg$ will depend on these quantities only. As before, whenever we speak of the height of a polynomial it will be understood that its coefficients are integers. We shall denote by $f^{(j)}$ the jth derivative of a function f, or f' in the case of the first derivative.

Lemma 1. *For any positive integers $r_1, \ldots, r_n$ with maximum $r \gg 1$, there exist polynomials $P_i(x)$ $(1 \leqslant i \leqslant n)$, not all identically 0, with degrees at most r and heights at most $r_i!\,r^{\epsilon r}$, such that $P_i^{(j)}(0) = 0$ for*

† *V.M.* 2 (1967), 63–72.
‡ Cf. *Ann. Univ. Sci. Budapest*, 9 (1966), 3–14 (Luise-Charlotte Kappe).
§ *Michigan Math. J.* 11 (1964), 247–50.

$j < r - r_i$, *and*

$$\sum_{i=1}^{n} P_i(x)\, e^{\theta_i x} = \sum_{m=M}^{\infty} \rho_m x^m, \tag{1}$$

where $|\rho_m| < (r!/m!)\, r^{\epsilon(r+m)}$ *and*

$$M = r_1 + \ldots + r_n + n - 1 - [\epsilon r].$$

Proof. Let L be the maximum of the absolute values of $\theta_1, \ldots, \theta_n$ and let l be the least common multiple of their denominators. We take p_{ij} to be 0 for all integers i, j other than the $N = r_1 + \ldots + r_n + n$ pairs given by $1 \leqslant i \leqslant n$ and $r - r_i \leqslant j \leqslant r$, and we then define p_{ij} for these remaining values as integers, not all 0, satisfying

$$\sum_{i=1}^{n} \sum_{j=0}^{m} \binom{m}{j} \theta_i^{m-j} l^m p_{ij} = 0 \quad (0 \leqslant m < M). \tag{2}$$

Such integers exist by virtue of Lemma 1 of Chapter 2, and indeed they can be selected to have absolute values at most

$$H = \{N(2lL)^M\}^{M/(N-M)}.$$

We proceed to prove that the polynomials

$$P_i(x) = r! \sum_{j=0}^{r} p_{ij} (j!)^{-1} x^j \quad (1 \leqslant i \leqslant n)$$

have the required properties.

First we observe that, on expanding $e^{\theta_i x}$ as a power series in x, we obtain

$$\sum_{i=1}^{n} P_i(x)\, e^{\theta_i x} = r! \sum_{m=0}^{\infty} \sigma_m (m!)^{-1} x^m,$$

where, for each m, $l^m \sigma_m$ is given by the left-hand side of (2). Hence (1) holds with $\rho_m = (r!/m!)\, \sigma_m$. Further we have $M < N < 2nr$ and $N - M > \epsilon r$, whence

$$H < \{2nr(2lL)^{2nr}\}^{2n/\epsilon} < r^{\frac{1}{2}\epsilon r}.$$

Since $p_{ij} = 0$ for $j < r - r_i$ it follows that the coefficients of the $P_i(x)$ have absolute values at most

$$\frac{r!\,H}{(r - r_i)!} = r_i!\, H \binom{r}{r_i} \leqslant r_i!\, r^{\epsilon r}.$$

Also it is clear that

$$|\sigma_m| < n(m+1)\,(2lL)^m H < r^{\epsilon(r+m)},$$

and this proves the lemma.

Lemma 2. *Let* $P_{ij}(x)$ $(1 \leqslant i \leqslant n, j \geqslant 1)$ *be defined recursively by*

$$P_{i1}(x) = P_i(x), \quad P_{i,j+1}(x) = P'_{ij}(x) + \theta_i P_{ij}(x).$$

If $r_i > 2s$ for all i, where $s = [\epsilon r] + (n-1)^2$, then the determinant $\Delta(x)$ of order n with $P_{ij}(x)$ in the ith row and jth column cannot have a zero at $x = 1$ with order greater than s.

Proof. We shall show in a moment that none of the $P_i(x)$ is identically 0; at first we assume this. Then each P_i has a non-zero leading coefficient p_i say. Since clearly $P_{ij}(x)$ has degree at most r and leading coefficient $p_i\theta_i^{j-1}$, it follows that $\Delta(x)$ is a polynomial with degree at most nr and with leading coefficient $p_1 \dots p_n \psi$, where ψ is a Vandermonde determinant of order n formed from the powers of the θ_i. By hypothesis, the θ_i are distinct and so $\Delta(x)$ is not identically 0.

We suppose now, as we may without loss of generality, that $r = r_1$. Denoting the left-hand side of (1) by $\Phi(x)$, we clearly have

$$\Phi^{(j-1)}(x) = \sum_{i=1}^{n} P_{ij}(x)\, e^{\theta_i x}.$$

Hence $\Delta(x)$ remains unaltered if the first row is replaced by $e^{-\theta_1 x}\Phi^{(j-1)}(x)$ with $j = 1, 2, \dots, n$. On differentiating (1), we see that $\Phi^{(j)}(x)$ has a zero at $x = 0$ with order at least $M - j$; and clearly $P_{ij}(x)$ has a zero at $x = 0$ with order at least $r - r_i - j + 1$. Hence $\Delta(x)$ has a zero at $x = 0$ with order at least

$$M - n + 1 + \sum_{i=2}^{n} (r - r_i - n + 1) = nr - s,$$

and the lemma follows since $\Delta(x)$ has degree at most nr.

It remains only to prove the original supposition. We suppose that exactly k of the polynomials $P_i(x)$ do not vanish identically and, without loss of generality, that these are given by $i = 1, 2, \dots, k$. Also we assume, as clearly we may, that $r = r_i$ for some i with $k \leqslant i \leqslant n$. Now, as above, we see that the minor in $\Delta(x)$ formed from the first k rows and columns is a polynomial, not identically 0, with degree at most kr. On the other hand, on taking a linear combination of rows, it is clear that it has a zero at $x = 0$ with order at least

$$M - k + 1 + \sum_{i=1}^{k-1} (r - r_i - k + 1) \geqslant (k-1)r - s + \sum_{i=k}^{n} r_i.$$

By virtue of the hypothesis $r_i > 2s$ for all i, it follows that $k = n$, and this completes the proof of the lemma.

Lemma 3. *There are n distinct suffixes $J(j)$ $(1 \leqslant j \leqslant n)$ between 1 and $n + s$ inclusive such that the determinant of order n with $P_{i,J(j)}(x)$ in the ith row and jth column does not vanish at $x = 1$.*

Proof. We introduce linear forms in $w_1, \ldots, w_n$ by the equations

$$W_j = \sum_{i=1}^{n} P_{ij}(x)\, w_i \quad (j = 1, 2, \ldots). \tag{3}$$

If $\Delta_{ij}(x)$ is the minor in $\Delta(x)$ formed by omitting the ith row and jth column then

$$w_i \Delta(x) = \sum_{j=1}^{n} (-1)^{i+j} W_j \Delta_{ij}(x) \quad (1 \leqslant i \leqslant n). \tag{4}$$

By Lemma 2, there is an integer $t \leqslant s$ such that $\Delta^{(t)}(1) \neq 0$ and we suppose that t is the least such non-negative integer. Now regarding the w_j as differentiable functions of x and differentiating (4) t times, replacing the w_i' occurring at each stage by $w_i\theta_i$ (as we may since the resulting equations hold identically in the w_i and w_i') we obtain

$$w_i\left\{\sum_{j=0}^{t} \binom{t}{j} \theta_i^{t-j} \Delta^{(j)}(x)\right\} = \sum_{j=1}^{n+t} W_j F_{ij}(x) \quad (1 \leqslant i \leqslant n),$$

where the $F_{ij}(x)$ are polynomials given by linear combinations of the $\Delta_{ij}(x)$ and their derivatives. Hence the linear forms defined by (3) with $x = 1$ and with $1 \leqslant j \leqslant n+t$, include a set of n linearly independent forms, and the lemma follows with $J(j)$ $(1 \leqslant j \leqslant n)$ given by the associated suffixes.

Lemma 4. *There are integers q_{ij} $(1 \leqslant i, j \leqslant n)$, forming a non-zero determinant, such that $|q_{ij}| < r_i!\, r^{4cr}$ and*

$$\left|\sum_{i=1}^{n} q_{ij} e^{\theta_i}\right| < r!\, r^{4cnr} \left(\prod_{i=1}^{n} r_i!\right)^{-1}. \tag{5}$$

Proof. In fact the integers $q_{ij} = l^{n+s} P_{i,J(j)}(1)$ have the required properties. Indeed the first assertion follows from Lemma 3 and the second from the obvious upper estimate $r_i!\,(r+L)^j r^{cr}$ for the absolute values of the coefficients of P_{ij}. Further, with the notation of Lemma 2, the sum on the left of (5) is given by $l^{n+s}\Phi^{J(j)-1}(1)$, and, on differentiating (1) $h \leqslant n+s-1$ times, we obtain

$$|\Phi^{(h)}(1)| < r!\, r^{cr} \sum_{m=M}^{\infty} r^{cm}((m-h)!)^{-1}.$$

But the sum on the right multiplied by $(M-h)!$ is clearly at most $e^r r^{cM}$, and we have

$$(M-h)! \geqslant (r_1 + \ldots + r_n - 2s)! \geqslant (nr)^{-2s}(r_1 + \ldots + r_n)!.$$

Since $M \leqslant \frac{5}{4}nr$ and $s \leqslant \frac{5}{4}cr$, this gives (5).

3. Proof of main theorem

The proof can now be completed readily by means of the Geometry of Numbers.† Let $\theta_1, \ldots, \theta_n$ be distinct non-zero rationals and suppose that $\xi > 0$. Constants implied by $\ll$ or $\gg$ will depend on these quantities only. For brevity we put $k = n+1$, and we signify by $\mathbf{A}_k$ the vector $(e^{\theta_1}, \ldots, e^{\theta_n}, 1)$ in R^k. Further we signify by $\mathbf{A}_j$ $(1 \leqslant j \leqslant n)$ the jth row of the unit matrix of order k. We proceed to show that, for any numbers $\mu_1, \ldots, \mu_k$ with $\mu_1 \ldots \mu_k = 1$ and $\mu_j > 1$ $(1 \leqslant j < k)$, the first minimum λ_1 of the parallelepiped $|\mathbf{A}_j \mathbf{x}| < \mu_j$ $(1 \leqslant j \leqslant k)$ exceeds $\mu^{-\xi}$ if $\mu_j \ll \mu$ for all j and $\mu \gg 1$.

In fact it suffices to show that the last minimum λ_k of the parallelepiped is $\ll \mu^{\xi/n}$, for we have $\lambda_1 \ldots \lambda_k \gg 1$ and so $\lambda_1 \gg \lambda_k^{-n}$. We shall apply Lemma 4 with n replaced by k and with $\theta_k = 0$. We take $r = r_k$ to be the least positive integer for which $\mu < r!\,r^{-4\epsilon r}$, and we then take $r_1, \ldots, r_n$ to be the integers satisfying

$$(r_j - 1)! \leqslant \mu_j r^{4\epsilon r} < r_j!.$$

Clearly r is the maximum of $r_1, \ldots, r_k$ and we have $r \gg 1$ and $r_i > 4\epsilon r$ for all i; in particular, the hypothesis of Lemma 2 is satisfied. Further, from Stirling's formula we see that

$$\mu > (r-1)!\,r^{-4\epsilon r} > r^{\frac{1}{2}r},$$

and so, by Lemma 4,

$$|q_{ij}| < \mu_i r^{8\epsilon r+1} < \mu_i \mu^{20\epsilon}.$$

Further, the right-hand side of (5) is at most

$$r^{4\epsilon r}(\mu_1 \ldots \mu_n)^{-1} \leqslant \mu_k \mu^{20\epsilon},$$

and since the determinant of the q_{ij} is not 0, it follows that $\lambda_k \leqslant \mu^{20\epsilon}$. This gives $\lambda_1 > \mu^{-\xi}$ if ϵ is sufficiently small, as required.

Finally, we apply Lemma 8 of Chapter 7 with $l = n = k-1$. Denoting by $\mathbf{a}_j$ $(1 \leqslant j \leqslant k)$ the vectors defined at the beginning of § 10 of Chapter 7 with e^{θ_j} in place of α_j, we conclude that the first minimum ν_1 of the parallelepiped $|\mathbf{a}_j \mathbf{x}| < \mu_j^{-1}$ $(1 \leqslant j \leqslant k)$ satisfies

$$\nu_1 \gg \lambda_1 \ldots \lambda_l \geqslant \lambda_1^l$$

and so $\nu_1 \gg \mu^{-l\xi}$. Hence the main proposition of § 10 holds, and Theorem 10.1 now follows by the argument immediately succeeding.

† For an alternative argument see the author's memoir in *Canadian J. Math.* **17** (1965).

11

THE SIEGEL–SHIDLOVSKY THEOREMS

1. Introduction

In 1929 Siegel† obtained a general method for establishing the algebraic independence of the values of a certain class of power series satisfying systems of linear differential equations. Siegel called the power series in question E-functions. By this he meant series of the form

$$\sum_{n=0}^{\infty} a_n x^n/n!,$$

with $a_0, a_1, \ldots$ elements of an algebraic number field such that, for some sequence $b_0, b_1, \ldots$ of positive integers and for any $\epsilon > 0$, $b_n a_0, \ldots, b_n a_n$ and b_n are all algebraic integers with size $\ll n^{\epsilon n}$, where the implied constant depends only on ϵ; here the size denotes, as in Chapter 4, the maximum of the absolute values of the conjugates. It is clear that the exponential function is an E-function, and indeed so is the normalized Bessel function

$$K_\lambda(x) = \Gamma(\lambda+1)\,(\tfrac{1}{2}x)^{-\lambda} J_\lambda(x) = \sum_{n=0}^{\infty} \frac{(-1)^n\,(\frac{1}{2}x)^{2n}}{n!\,(\lambda+1)\ldots(\lambda+n)}$$

for all rational values of λ other than the negative integers. More generally, any hypergeometric function

$$\sum_{n=0}^{\infty} \frac{[\alpha_1, n]\ldots[\alpha_l, n]}{[\beta_1, n]\ldots[\beta_m, n]} x^{kn}$$

is an E-function, where $k = m - l > 0$,

$$[\gamma, n] = \gamma(\gamma+1)\ldots(\gamma+n-1),$$

and the α's and β's are rationals other than negative integers. The latter assertion follows in fact from the observation that, for any rational $\alpha = p/q$, the integer $q^n[\alpha, n]$ divides $n!\,\nu$, where ν denotes the least common multiple of all the positive integers up to

$$m = (|p| + |q|)\,n;$$

and from the prime-number theorem we have $\nu < c^m$ for some absolute

† *Abh. Preuss. Akad. Wiss.* 1929, No. 1.

constant c. Furthermore, it is readily verified that sums, products, derivatives and integrals of E-functions are again E-functions.

Siegel's work related to differential equations of the first and second orders only, and it was an outstanding question for many years to devise a means of extending the arguments to higher order equations. The problem was solved by Shidlovsky† in 1954 and many notable applications have followed.‡ The basic result concerns E-functions $E_1(x), \ldots, E_n(x)$ satisfying a system of homogeneous linear differential equations

$$y_i' = \sum_{j=1}^{n} f_{ij}(x)\, y_j \quad (1 \leqslant i \leqslant n), \tag{1}$$

where the f_{ij} are rational functions of x, and the coefficients of all the E's and f's are supposed to be elements of an algebraic number field K. We have then

Theorem 11.1. *If $E_1(x), \ldots, E_n(x)$ are algebraically independent over $K(x)$ then, for any non-zero algebraic number α distinct from the poles of the f_{ij}, $E_1(\alpha), \ldots, E_n(\alpha)$ are algebraically independent.*

The theorem can easily be extended to yield an assertion to the effect that the maximum number of algebraically independent elements among $E_1(x), \ldots, E_n(x)$ is the same as that among

$$E_1(\alpha), \ldots, E_n(\alpha),$$

and moreover there is no difficulty in generalizing the latter result to inhomogeneous equations where an additional rational function is present on the right of (1). As an immediate application of Theorem 11.1, we see that if λ is rational, but not a negative integer or half an odd integer, then $K_\lambda(\alpha)$ and $K_\lambda'(\alpha)$ are algebraically independent for every non-zero algebraic number α; for it is well known§ that $K_\lambda(x)$ and $K_\lambda'(x)$ are algebraically independent over $Q(x)$. This further implies, for example, that the continued fraction with partial quotients $1, 2, 3, \ldots$ is transcendental; for $J_0(\sqrt{(-4x)})$ $[= K_0(\sqrt{(-4x)})]$ satisfies the differential equation $xy'' + y' = y$, and the continued fraction in question is given by y/y' evaluated at $x = 1$. Oleinikov‖ has obtained some similar theorems for third order linear differential equations; for instance he has shown that if

$$F(x) = \sum_{n=0}^{\infty} \frac{(x/3)^{3n}}{n!\,[\lambda, n]\,[\mu, n]},$$

† *I.A.N.* **23** (1959), 35–66.
‡ Cf. the survey of Feldman and Shidlovsky (Bibliography).
§ Cf. Siegel (Bibliography).
‖ *D.A.N.* **166** (1966), 540–3.

where λ, μ are rationals such that none of $\lambda+\mu$, $\lambda-2\mu$, $\mu-2\lambda$ are integers, then $F(x)$, $F'(x)$, $F''(x)$ satisfy the hypothesis of Theorem 11.1, whence $F(\alpha)$, $F'(\alpha)$, $F''(\alpha)$ are algebraically independent for every non-zero algebraic number α. And Shidlovsky† has proved a striking theorem to the effect that if

$$\Phi_k(x) = \sum_{n=0}^{\infty} x^{kn}/(n!)^k,$$

then, for any non-zero algebraic α and any r, the numbers $\Phi_k^{(l)}(\alpha/k)$, with $1 \leqslant l < k$, $1 \leqslant k \leqslant r$, are algebraically independent. Plainly also Theorem 11.1 includes Lindemann's theorem.

2. Basic construction

The proof of Theorem 11.1 follows closely the arguments of the preceding chapter, but it is no longer a simple matter to confirm that $\Delta(x)$ does not vanish identically. The verification, which is Shidlovsky's major discovery in the subject, will be given in Lemma 2 below.

We shall signify by $E_1(x), \ldots, E_n(x)$ E-functions as above, linearly independent over $K(x)$ and we shall suppose that $0 < \epsilon < 1$. Constants implied by $\ll$ or $\gg$ will depend on the coefficients in the E's, f's and on ϵ only. By $f(x)$ we signify a polynomial, not identically 0, with coefficients in K, such that ff_{ij} is a polynomial for all f_{ij} in (1).

Lemma 1. *For any integer $r \gg 1$, there are polynomials*

$$P_i(x) \quad (1 \leqslant i \leqslant n),$$

not all identically 0, *with degrees at most r and algebraic integer coefficients in K with sizes at most $(r!)^{1+\epsilon}$, such that*

$$\sum_{i=1}^{n} P_i(x)\, E_i(x) = \sum_{m=M}^{\infty} \rho_m x^m, \tag{2}$$

where $|\rho_m| < r!\,(m!)^{-1+\epsilon}$ and

$$M = n(r+1)-1-[\epsilon r].$$

Proof. Let a_{ij} be the coefficient of $x^j/j!$ in $E_i(x)$ and let $b_{i0}, b_{i1}, \ldots$ be the sequence of integers associated with E_i as in § 1. By Lemma 1 of Chapter 6, there exist algebraic integers p_{ij} $(1 \leqslant i \leqslant n,\ 0 \leqslant j \leqslant r)$ in K,

† *Trudy Moskov.* **18** (1968), 55–64.

not all 0, such that

$$\sum_{i=1}^{n} \sum_{j=0}^{\min(r,m)} \binom{m}{j} a_{i,m-j} p_{ij} = 0 \quad (0 \leqslant m < M), \tag{3}$$

and indeed they can be selected to have sizes at most $N^{\delta NM/(N-M)}$, where $N = n(r+1)$ and $\delta = (\epsilon/4n)^2$; for, on multiplying (3) by $b_{1m} \ldots b_{nm}$, the coefficients become algebraic integers in K with sizes $\ll 2^M M^{\frac{1}{2}\delta M}$, as is clear on taking $\delta/(2n)$ in place of ϵ in the defining property of the b's. We conclude, as in the proof of Lemma 1 of Chapter 10, that the polynomials

$$P_i(x) = r! \sum_{j=0}^{r} p_{ij}(j!)^{-1} x^j \quad (1 \leqslant i \leqslant n)$$

have the asserted properties. In fact (2) plainly holds with

$$\rho_m = (r!/m!)\, \sigma_m,$$

where σ_m denotes the left-hand side of (3), and since $M < N < 2nr$ and $N - M > \epsilon r$, we see that the p_{ij} have sizes at most $r^{\frac{1}{2}\epsilon r}$, whence $|\sigma_m| < (m!)^\epsilon$ for $m \geqslant M$, as required.

Lemma 2. *Let $P_{ij}(x)$ $(1 \leqslant i \leqslant n, j \geqslant 1)$ be defined recursively by*

$$P_{i1} = P_i, \quad P_{i,j+1} = f P'_{ij} + f \sum_{h=1}^{n} f_{hi} P_{hj}.$$

Then the determinant $\Delta(x)$ of order n with $P_{ij}(x)$ in the ith row and jth column is not identically 0.

Proof. Suppose, on the contrary, that $\Delta(x)$ vanishes identically. Let k be the integer such that the first k columns of $\Delta(x)$ are linearly independent over $K(x)$ but the $(k+1)$th column is linearly dependent on these. We signify by $\mathbf{Q}$ the matrix formed by the first k columns of $\Delta(x)$, and by $\mathbf{R}$ and $\mathbf{S}$ the matrices formed from the first k rows of $\mathbf{Q}$ and last $n-k$ rows respectively. We assume, as clearly we may, that the notation is such that $\mathbf{R}$ is non-singular, and we proceed to prove that the degrees of the numerators and denominators of the rational function elements of $\mathbf{SR}^{-1}$ are $\ll 1$, where in fact the implied constant depends only on the f's. This will suffice to establish the lemma; for denoting by $\mathbf{L}$ the row vector with jth element

$$L_j = \sum_{i=1}^{n} P_{ij} E_i \quad (1 \leqslant j \leqslant k),$$

and putting $\mathbf{A} = (E_1, \ldots, E_k)$, $\mathbf{B} = (E_{k+1}, \ldots, E_n)$,

we have $\mathbf{L} = \mathbf{AR} + \mathbf{BS}$ whence

$$\mathbf{LR}^{-1} = \mathbf{A} + \mathbf{BSR}^{-1}. \tag{4}$$

But L_j satisfies the differential equation $L_{j+1} = fL'_j$ and so each element of $\mathbf{L}$ has a zero at $x = 0$ with order at least $M-n$. Further, each element of $\mathbf{R}^{-1}$ can be expressed as a rational function in $K(x)$ with denominator $\det \mathbf{R}$, and since the latter is a polynomial with degree at most $kr+c$, where $c \ll 1$, it follows that each element of $\mathbf{LR}^{-1}$ has a zero at $x = 0$ with order at least $M-kr-n-c$. On the other hand, the vector on the right of (4) cannot vanish identically in view of the assumed linear independence of $E_1, \ldots, E_n$, and the order of the zeros of its elements at $x = 0$, if any, are bounded independently of the coefficients of the elements of $\mathbf{SR}^{-1}$, and so, in particular, of r. Now $k < n$, and so $M-kr$ tends to infinity with r; hence we have a contradiction if r is sufficiently large.

To prove the assertion concerning $\mathbf{SR}^{-1}$, we observe first that there is a square matrix $\mathbf{F}$ of order k, with elements in $K(x)$, such that, for any solution $\mathbf{y}$ of (1), the vector $\mathbf{Y} = \mathbf{yQ}$ satisfies the differential equation $\mathbf{Y}' = \mathbf{YF}$. Indeed if Y_j denotes the jth element of $\mathbf{Y}$, then $Y_{j+1} = fY'_j$ for all $j < k$ and, by the definition of k, fY'_k is expressible as a linear combination of $Y_1, \ldots, Y_k$ with coefficients in $K(x)$. Let now $\mathbf{w}_1, \ldots, \mathbf{w}_n$ be power series solutions of (1) linearly independent over K and let $\mathbf{W}$ be the square matrix of order n with jth row $\mathbf{w}_j$. Then each row of $\mathbf{WQ}$ is a solution of $\mathbf{Y}' = \mathbf{YF}$; but this has at most k solutions linearly independent over K and thus there exists an $n-k$ by n matrix $\mathbf{M}$ with coefficients in K and rank $n-k$ satisfying $\mathbf{MWQ} = 0$. Denoting by $\mathbf{U}$ and $\mathbf{V}$ the matrices formed from the first k columns of $\mathbf{MW}$ and the last $n-k$ columns respectively, we have $\mathbf{UR} + \mathbf{VS} = 0$. Since $\mathbf{R}$ is non-singular and $\mathbf{MW}$ has rank $n-k$ it follows that $\mathbf{V}$ is non-singular and so $\mathbf{SR}^{-1} = -\mathbf{V}^{-1}\mathbf{U}$. Clearly the elements of $\mathbf{V}^{-1}\mathbf{U}$ are rational functions in the elements of $\mathbf{W}$ with coefficients in K and with the degrees of the numerators and denominators bounded independently of r. Hence they can be expressed as quotients of linear forms in certain monomials in the elements of $\mathbf{W}$, linearly independent over $K(x)$, the coefficients in the linear forms being rational functions in $K(x)$ for which again the degrees of the numerators and denominators are bounded independently of r. Since the elements of $\mathbf{SR}^{-1}$ and so also of $\mathbf{V}^{-1}\mathbf{U}$ are in fact in $K(x)$, they must be given by quotients of such coefficients, and the assertion follows.

3. Further lemmas

We now obtain analogues of Lemmas 3 and 4 of Chapter 10. The arguments here will follow closely their earlier counterparts and so we shall be relatively brief.

By α we shall signify an element of K with $\alpha f(\alpha) \neq 0$. By $c_1, c_2, \ldots$ we denote positive numbers which may depend on α, ϵ and the coefficients in the E's and f's only.

Lemma 3. *There are distinct suffixes $J(j)$ $(1 \leqslant j \leqslant n)$ not exceeding $\epsilon r + c_1$ such that the determinant with $P_{i, J(j)}(x)$ in the ith row and jth column does not vanish at $x = \alpha$.*

Proof. We begin by noting that $\Delta(x)$ remains unaltered if the first row is replaced by $E_1^{-1} L_j$ with $j = 1, 2, \ldots, n$, where L_j is defined as in the proof of Lemma 2. Hence $\Delta(x)$ has a zero at $x = 0$ with order at least $M - c_2$, and since it is a polynomial with degree at most $nr + c_3$, it follows that a non-negative integer t exists, not exceeding

$$nr + c_3 - (M - c_2) \leqslant \epsilon r + c_4,$$

such that $\Delta^{(t)}(\alpha) \neq 0$; we suppose that t is chosen minimally.

We now introduce linear forms in $w_1, \ldots, w_n$ by (3) of Chapter 10. On applying the operator fd/dx to (4) of that chapter t times, replacing w_i' occurring at each stage by the right-hand side of (1) with $y_j = w_j$, we obtain

$$w_i(f(\alpha))^t \Delta^{(t)}(\alpha) = \sum_{j=1}^{n+t} W_j F_{ij}(\alpha) \quad (1 \leqslant i \leqslant n),$$

where the F_{ij} denote polynomials in x given by linear combinations of the f's, Δ's and their derivatives. Hence the linear forms

$$W_j \quad (1 \leqslant j \leqslant n+t)$$

with $x = \alpha$ include a set of n linearly independent forms and the lemma follows with $J(j)$ given by the associated suffixes.

Lemma 4. *There are algebraic integers q_{ij} $(1 \leqslant i, j \leqslant n)$ in K with sizes at most $(r!)^{1+16\epsilon}$ forming a non-zero determinant and satisfying*

$$\left|\sum_{i=1}^{n} q_{ij} E_i(\alpha)\right| < (r!)^{-n+1+16\epsilon n} \quad (1 \leqslant j \leqslant n). \tag{5}$$

Proof. Let l be a positive integer such that $l\alpha$ and the coefficients in

lf and all lff_{ij} are algebraic integers. We proceed to prove that the numbers

$$q_{ij} = l^{r+(m+1)J(j)}P_{i,J(j)}(\alpha) \quad (1 \leqslant i, j \leqslant n),$$

where m denotes the maximum of the degrees of the ff_{ij} and f, have the required properties. First it is clear that $l^j P_{ij}$ has algebraic integer coefficients and degree at most $r+mj$. Thus the q's are algebraic integers and, by Lemma 3, they form a non-zero determinant. Further, it is easily verified by induction that the sizes of the coefficients of $l^j P_{ij}$ are at most $(r+mj)^{2j} c_6^j (r!)^{1+\epsilon}$, and since the $J(j)$ do not exceed $\epsilon r + c_1$, this gives the required estimate for the sizes of the q's.

It remains to prove (5). Denoting by $\Phi(x)$ the left-hand side of (2), it is clear that the sum on the left of (5) is given, apart from a factor $l^{r+(m+1)J}$, by $(fd/dx)^{J-1}\Phi$ evaluated at $x=\alpha$, where $J = J(j)$. But, again by induction, we see that this is a linear form in the $\Phi^{(h)}(\alpha)$, where $h = 0, 1, \ldots, J-1$, having coefficients with absolute values at most $(c_6 J)^{2J}$. Hence it suffices to prove that

$$|\Phi^{(h)}(\alpha)| < (r!)^{-n+1+8\epsilon n} \quad (0 \leqslant h < J).$$

Now from Lemma 1 we obtain

$$|\Phi^{(h)}(\alpha)| < r! \sum_{m=M}^{\infty} (m!)^{\epsilon} ((m-h)!)^{-1} |\alpha|^{m-h},$$

and the sum on the right is at most

$$h! \sum_{m=M}^{\infty} (m!)^{-1+\epsilon} 2^m |\alpha|^{m-h} \leqslant h!\, c_7^M (M!)^{-1+\epsilon}.$$

Since $h < \epsilon r + c_1$ and $M \leqslant 2nr$ we have $h! \leqslant (r!)^{3\epsilon}$ and

$$M! \geqslant (2nr)^{-\epsilon r} (r!)^n \geqslant (r!)^{n-3\epsilon}.$$

The required estimate follows at once.

4. Proof of main theorem

Suppose that $E_1(\alpha), \ldots, E_n(\alpha)$ are algebraically dependent. Then they satisfy an equation $P(E_1, \ldots, E_n) = 0$, where P is a polynomial with algebraic coefficients, not all 0. We shall denote by c the degree of P, and we shall assume, as we may without loss of generality, that the coefficients in P are algebraic integers in K. The degree of K will be denoted by d, and we shall suppose that $0 < \epsilon < 1$. Further we shall signify by m an integer such that the binomial coefficients

$$k = \binom{m+n+c}{n}, \quad l = \binom{m+n}{n}$$

satisfy $k-l < l/(2d)$; the latter inequality certainly holds for all sufficiently large m since k and l are asymptotic to $m^n/n!$ as $m \to \infty$, as is easily seen by expressing them as polynomials in m with degree n. In the sequel, constants implied by $\ll$ or $\gg$ will depend on α, ϵ, m and the coefficients in the E's, f's and P only.

Let now $\mathscr{E}_1, \ldots, \mathscr{E}_k$ be the E-functions $E_1^{j_1} \ldots E_n^{j_n}$, where $j_1, \ldots, j_n$ run through all non-negative integers with $j_1 + \ldots + j_n \leqslant m + c$. Then clearly $\mathscr{E}_1, \ldots, \mathscr{E}_k$ satisfy a further system of linear differential equations of the form (1), where the new coefficients are given by linear combinations of the f's; furthermore, $\mathscr{E}_1, \ldots, \mathscr{E}_k$ are linearly independent over $K(x)$ by virtue of the hypothesis regarding the algebraic independence of $E_1(x), \ldots, E_n(x)$. We conclude from § 2 and § 3 that, for any integer $r \gg 1$, there exist algebraic integers q_{ij} $(1 \leqslant i, j \leqslant k)$ in K possessing the properties cited in Lemma 4 with $\mathscr{E}_1, \ldots, \mathscr{E}_k$ in place of $E_1, \ldots, E_n$. For each set of non-negative integers $j_1, \ldots, j_n$ with $j_1 + \ldots + j_n \leqslant m$ we write

$$E_1^{j_1} \ldots E_n^{j_n} P(E_1, \ldots, E_n) = \sum_{i=1}^{k} p_{ij} \mathscr{E}_i,$$

where the p_{ij} are either coefficients in P or 0, and $j = j(j_1, \ldots, j_n)$ takes the values $1, 2, \ldots, l$. Then on the right we have l linear forms in $\mathscr{E}_1, \ldots, \mathscr{E}_k$ linearly independent over K, all of which vanish at $x = \alpha$. Since the determinant of the q_{ij} is not 0, it follows that there exist $k-l$ of the forms

$$\Phi_j = \sum_{i=1}^{k} q_{ij} \mathscr{E}_i \quad (1 \leqslant j \leqslant k),$$

which together with the latter make up a linearly independent set; without loss of generality we can suppose that they are given by $\Phi_{l+1}, \ldots, \Phi_k$. We shall suppose also, as clearly we may, that $\mathscr{E}_1(\alpha) \neq 0$.

We now compare estimates for the determinant D of order k with p_{ij} in the ith row and jth column for $j \leqslant l$ and q_{ij} in that position for $j > l$. Plainly D is a non-zero algebraic integer in K, and, since $p_{ij} \ll 1$, it has size $\ll (r!)^{(1+16\epsilon)(k-l)}$; hence

$$|D| \gg (r!)^{-(1+16\epsilon)(k-l)d} \geqslant (r!)^{-(1+16\epsilon)l/2}.$$

On the other hand, D is unaltered if the first row is replaced by 0 for $j \leqslant l$ and by $\mathscr{E}_1^{-1}(\alpha)\,\Phi_j$ for $j > l$. Further, by Lemma 4, the latter elements are $\ll (r!)^{-k+1+16\epsilon k}$; thus

$$|D| \ll (r!)^{(1+16\epsilon)(k-l-1)-k+1+16\epsilon k} \leqslant (r!)^{-l+32\epsilon k}.$$

But $k < \frac{3}{2}l$ and so, if $\epsilon < \frac{1}{112}$ and r is sufficiently large, we have a contradiction. This proves the theorem.

Subsequent to the fundamental discovery of Shidlovsky, researches in this field have largely centred on establishing the function–theoretic hypotheses of Theorem 11.1 and its extensions for particular classes of E-functions, and, as indicated in § 1, this has in fact been accomplished in many striking cases. Studies have also been carried out in connexion with obtaining positive lower bounds for expressions of the type $P(E_1, \ldots, E_n)$ as above, and in fact an estimate of the form Ch^{-c} has been established, where h denotes the maximum of the sizes of the coefficients of P and C, c are positive numbers which do not depend on h; but c here increases rapidly with n.† The main outstanding problem in the subject is to generalize the theory to wider classes of analytic functions than E-functions, and any progress here would be of much interest.

† Cf. Lang (Bibliography, first work).

12

ALGEBRAIC INDEPENDENCE

1. Introduction

Few theorems have been established to date on algebraic, as opposed to linear, independence of transcendental numbers. Indeed, apart from the results on E-functions discussed in the last chapter, which in fact follow at once from their linear analogues, and the examples mentioned in Chapter 8 that arise from the properties of Mahler's classification, the only work in this context of a general nature is based on studies of Gelfond[†] carried out in 1949. Recently a number of authors have obtained important improvements in this connexion, and these latest developments will be the theme of the present chapter.

The essential character of the results is well-illustrated by:

Theorem 12.1. *If both ξ_1, ξ_2, ξ_3 and η_1, η_2, η_3 are linearly independent over the rationals, then two at least of the numbers*

$$\xi_i, \quad e^{\xi_i \eta_j} \quad (1 \leqslant i, j \leqslant 3)$$

are algebraically independent.

Gelfond proved the theorem originally subject to certain supplementary conditions, and the formulation here is due to Tijdeman.[‡] As an immediate consequence one sees that if α is an algebraic number other than 0 or 1 and β is a cubic irrational then $\alpha^\beta, \alpha^{\beta^2}$ are algebraically independent; this follows in fact on taking $\xi_j = \beta^{j-1}$ and $\eta_j = \xi_j \log \alpha$. Tijdeman also derived two variants of Theorem 12.1; he proved that if $\xi_1, \xi_2, \xi_3, \xi_4$ and η_1, η_2 are linearly independent over the rationals, then two at least of $\xi_i, e^{\xi_i \eta_j}$ are algebraically independent, and moreover that if ξ_1, ξ_2, ξ_3 and η_1, η_2 are linearly independent over the rationals, then two at least of $\xi_i, \eta_j, e^{\xi_i \eta_j}$ are algebraically independent. These results include some earlier theorems of Šmelev.[§]

Very recently, Brownawell[‖] and Waldschmidt[¶] succeeded independently in obtaining a new version of the latter result which sufficed to solve a well-known problem of Schneider. They proved:

† Bibliography

‡ *I.M.* **33** (1971), 146–62.

§ *Mat. Zametki*, **3** (1968), 51–8; **4** (1968), 525–32.

‖ *J. Number Th.* **6** (1974), 11–31.

¶ *J. Number Th.* **5** (1973), 191–202.

Theorem 12.2. *If both ξ_1, ξ_2 and η_1, η_2 are linearly independent over the rationals and if $e^{\xi_1\eta_2}$ and $e^{\xi_2\eta_1}$ are algebraic, then two at least of ξ_i, η_j, $e^{\xi_i\eta_j}$ are algebraically independent.*

This implies, more especially, that if ξ_1, ξ_2 and η_1, η_2 are linearly independent over the rationals then at least two different numbers amongst ξ_i, η_j, $e^{\xi_i\eta_j}$ are transcendental. It follows at once, on taking $\xi_1 = \eta_1 = 1, \xi_2 = \eta_2 = e$, that one at least of e^e and e^{e^2} is transcendental. Furthermore, from Theorem 12.2, one sees, for instance, that at least one of $\alpha^{\log\alpha}$ and $\alpha^{(\log\alpha)^2}$ is transcendental for any algebraic number α other than 0 or 1. These results represent the nearest approach we have to date towards a confirmation of the transcendence of numbers of the type $\log\pi$ and e^{π^2}.

In another direction, Lang[†] has proved:

Theorem 12.3. *If ξ_1, ξ_2, ξ_3 and η_1, η_2 are linearly independent over the rationals then one at least of the numbers $e^{\xi_i\eta_j}$ is transcendental.*

Surprisingly, the demonstration of Theorem 12.3 is much simpler than that of Theorems 12.1 and 12.2, and yet the result admits several notable corollaries. In particular, it follows that, for any algebraic number α, not 0 or 1, and any transcendental β, one at least of α^β, α^{β^2}, α^{β^3} is transcendental; and in fact this result holds for any irrational β in view of the Gelfond–Schneider theorem. As a further example, the theorem plainly shows that for any real irrational β, the function x^β cannot assume algebraic values at more than two consecutive integral values of $x > 2$. More general results of this nature, involving, for instance, the Weierstrass $\wp$-function, were obtained by Ramachandra,[‡] who apparently discovered Theorem 12.3 independently. The theorem also throws some light on the problem raised by Schneider as to the untenability of the equation

$$\log\alpha \log\beta = \log\gamma \log\delta$$

in algebraic numbers $\alpha, \beta, \gamma, \delta$, having logarithms linearly independent over the rationals; it shows in fact that, given α, γ, there cannot be two solutions β, δ such that all six logarithms are linearly independent. The problem is, of course, only a special case of the wider open question as to a verification of the algebraic independence of the logarithms of algebraic numbers.

We remark finally that most of our expectations in connexion with

† Bibliography. ‡ *Acta Arith.* **14** (1968), 65–88.

the transcendence properties of the exponential and logarithmic functions are covered by a general conjecture, attributed to Schanuel, to the effect that if $\xi_1, \ldots, \xi_n$ are linearly independent over the rationals, then the transcendence degree of the field generated by $\xi_1, \ldots, \xi_n$, $e^{\xi_1}, \ldots, e^{\xi_n}$ over the rationals is at least n. The conjecture includes Theorems 1.4 and 2.1, and moreover it implies the algebraic independence of e and π. The power series analogue has been proved by Ax.†

2. Exponential polynomials

Our object here is to establish a theorem of Tijdeman‡ on the zeros of functions of the form

$$F(z) = \sum_{k=0}^{K-1} \sum_{l=1}^{L} f(k,l)\, z^k e^{\sigma_l z}.$$

We shall assume that $\sigma_1, \ldots, \sigma_L$ are complex numbers with absolute values at most S, and that the f's are arbitrary complex numbers for which F does not vanish identically. Constants implied by $\ll$ will be absolute. We prove:

Lemma 1. *The number of zeros of F in any closed disc, with radius R, counted with multiplicities, is $\ll KL + RS$.*

Tijdeman actually obtained the estimate $3KL + 4RS$, but the constants are not important for our purpose here. The main interest of the result lies in the fact that, in contrast to all previous theorems of its kind, there is no dependence on the differences between the σ's, and it is this strengthening that leads to the improvements in Gelfond's results mentioned earlier.

To commence the proof, let C be the circle centre the origin§ with radius R, and let $M(R)$ be the maximum of $|F|$ on C. Further, let

$$W(z) = (z-\omega_1) \ldots (z-\omega_h),$$

where $\omega_1, \ldots, \omega_h$ run through the zeros of F, taken with multiplicities, within and on C. Then F/W is regular within and on any concentric circle with larger radius, and so, by the maximum-modulus principle,

$$|W(v)|\, M(R) \leqslant |W(u)|\, M(4R),$$

where u, v are some numbers with $|u| = R$ and $|v| = 4R$. Now clearly

$$|W(u)| \leqslant (2R)^h, \quad |W(v)| \geqslant (3R)^h,$$

† *Ann. Math.* **93** (1971), 252–68.
‡ *I.M.* **33** (1971), 1–7.
§ Plainly, this choice involves no loss of generality.

and thus $$h \ll \log (M(4R)/M(R)).$$

It remains therefore to show that the number on the right is

$$\ll KL + RS.$$

Let the sequence $\sigma_1, \ldots, \sigma_1, \ldots, \sigma_L, \ldots, \sigma_L$ of $N = KL$ numbers, where each σ is repeated K times, be written as $\eta_1, \ldots, \eta_N$. By Newton's interpolation formula we have, for any w, z,

$$e^{zw} = \sum_{n=0}^{N} a_n P_n(w),$$

where $$P_n(w) = (w-\eta_1)\ldots(w-\eta_n) \quad (1 \leqslant n \leqslant N),$$

and $$a_n = \frac{1}{2\pi i}\int_\Gamma \frac{e^{z\zeta}}{P_{n+1}(\zeta)}\left(\frac{\zeta-\eta_{n+1}}{\zeta-w}\right)^{\delta_n} d\zeta,$$

Γ denoting a circle with centre the origin, described in the positive sense, including the η's and w, and $\delta_n = 0$ if $n < N$, $\delta_N = 1$. Clearly a_n is independent of w for $n < N$ and a_N is an integral function of w. We put

$$P(w) = \sum_{n=0}^{N-1} a_n P_n(w) = \sum_{n=0}^{N-1} p_n w^n,$$

and then it is readily verified that

$$F(z) = \sum_{k=0}^{K-1}\sum_{l=1}^{L} f(k,l)\, P^{(k)}(\sigma_l) = \sum_{n=0}^{N-1} p_n F^{(n)}(0).$$

We proceed now to employ the latter formula to obtain an upper bound for $|F|$.

By Cauchy's theorem we have

$$F^{(n)}(0) = \frac{n!}{2\pi i}\int_C \frac{F(\zeta)\,d\zeta}{\zeta^{n+1}},$$

and thus $$|F^{(n)}(0)| \leqslant n!\, M(R)/R^n.$$

This gives $$|F(z)| \leqslant M(R)\sum_{n=0}^{N-1} n!\,|p_n|/R^n.$$

To estimate the latter sum, let

$$b_n = \frac{1}{2\pi i}\int_\Gamma \frac{e^{|z|\zeta}}{Q_{n+1}(\zeta)}\,d\zeta,$$

where $Q_n(w) = (w-S)^n$ and Γ denotes a circle as above including S. On comparing the coefficients in $(P_n(\zeta))^{-1}$ and $(Q_n(\zeta))^{-1}$ when these are expressed as series in decreasing powers of ζ, we obtain $|a_n| \leqslant b_n$ for all $n < N$. But plainly

$$b_n = e^{|z|S}|z|^n/n!$$

and so, in view of the formula

$$n!\,p_n = \sum_{r=n}^{N-1} a_r P_r^{(n)}(0),$$

we have

$$n!\,|p_n| \leqslant n! \sum_{r=n}^{N-1} \binom{r}{n} S^{r-n} b_r = e^{|z|S} \sum_{r=n}^{N-1} |z|^r S^{r-n}/(r-n)! \leqslant |z|^n\, e^{2|z|S},$$

whence

$$|F(z)| \leqslant M(R)\, e^{2|z|S} \sum_{n=0}^{N-1} (|z|/R)^n.$$

On taking $|z| = 4R$, we conclude that

$$M(4R) \leqslant M(R)\, e^{8RS}\, 4^N,$$

and the lemma follows at once.

3. Heights

We shall require a more explicit version of Lemma 2 of Chapter 8. The result is due to Gelfond, who in fact obtained the proposition in a generalized form relating to polynomials in several variables.

Lemma 2. *If $P(x)$ is a polynomial with degree n and height h, and if $P = P_1 P_2 \ldots P_k$, where $P_j(x)$ is a polynomial with height h_j, then*

$$h \geqslant e^{-n} h_1 h_2 \ldots h_k.$$

We assume without loss of generality that $P(0) \neq 0$. For any zero ρ of P and any complex number z with $|z| = 1$, let w be the projection of ρ on the line through z and $-\rho/|\rho|$, taking $w = z$ if $z = \pm\rho/|\rho|$. Then, by simple geometry,

$$|z-\rho| \geqslant |w-\rho| = \tfrac{1}{2}(1+|\rho|)\,|z-\rho/|\rho||.$$

Thus, if $\rho_1, \ldots, \rho_n$ are all the zeros of P, then

$$|P(z)| \geqslant 2^{-n} M_1 \ldots M_k R(z),$$

where M_j denotes the maximum of P_j on the unit circle and

$$R(z) = \prod_{j=1}^{n} |z-\rho_j/|\rho_j||.$$

Now for any polynomial

$$Q(x) = q_0 + q_1 x + \dots + q_m x^m$$

we have
$$\int_0^1 |Q(e^{2\pi i\phi})|^2 \, d\phi = \sum_{j=0}^{m} |q_j|^2.$$

Hence taking $Q = R$ and noting that R has leading coefficient 1 and constant coefficient with absolute value 1, we obtain

$$\int_0^1 |P(e^{2\pi i\phi})|^2 \, d\phi \geq 2^{1-2n}(M_1 \dots M_k)^2.$$

But, on taking $Q = P$, we see that the number on the left is at most $2nh^2$, and clearly also

$$M_j^2 \geq \int_0^1 |P_j(e^{2\pi i\phi})|^2 \, d\phi \geq h_j^2.$$

Since $e^n \geq n^{\frac{1}{2}} 2^n$, this proves the lemma.

We shall require also a lemma closely related to the inequality $|\alpha - \beta| \gg a^{-l} b^{-m}$ mentioned in § 6 of Chapter 8. Again we shall adopt the convention that when one refers to the height of a polynomial it is implied that the coefficients are rational integers, not all 0.

Lemma 3. *If $P_1(x)$, $P_2(x)$ are polynomials with degrees n_1, n_2 and heights h_1, h_2 respectively and if P_1, P_2 have no common factor then, for any complex number z,*

$$\max(|P_1(z)|, |P_2(z)|) \geq (n_1 + n_2)^{-\frac{1}{2}(n_1+n_2+1)} h_1^{-n_2} h_2^{-n_1}.$$

The proof depends on the observation that since P_1, P_2 have no common factor, their resultant R is not 0. Now R can be expressed as the familiar Sylvester determinant of order $n_1 + n_2$ formed by eliminating x from the equations

$$x^i P_1(x) = 0 \quad (0 \leq i < n_2), \qquad x^j P_2(x) = 0 \quad (0 \leq j < n_1).$$

Thus R is a rational integer and so $|R| \geq 1$. On the other hand, R is unaltered if one replaces the element in the first column and ith row by $z^{i-1} P_1(z)$ for $i \leq n_2$ and by $z^{i-n_2-1} P_2(z)$ for $i > n_2$. Hence, if $|z| \leq 1$, the lemma follows from the upper estimates for the cofactors of these elements furnished by Hadamard's inequality. If $|z| > 1$ one argues similarly, replacing now the elements in the last column by numbers as above multiplied by $z^{-n_1-n_2+1}$.

4. Algebraic criterion

We now establish a lemma giving a sufficient condition for a number to be algebraic; it was derived in its original form by Gelfond and later sharpened by Brownawell and Waldschmidt. It shows that, in a sense, a number cannot be too well approximated by algebraic numbers unless it is itself algebraic and all the terms in the sequence beyond a certain point are equal. We shall actually prove the proposition in a form relating to polynomial sequences since this is more useful for applications.

First we need a preliminary lemma. Let $P(x)$ be a polynomial with degree n and height h, and let z be any complex number.

Lemma 4. *If $|P(z)| \leqslant 1$ then P has a factor Q, a power of an irreducible polynomial with integer coefficients, such that*

$$|Q(z)| \leqslant |P(z)| \exp(8n(n+\log h)).$$

We write P as a product $P_1 \dots P_k$ of powers of distinct irreducible polynomials and, for brevity, we put $p_j = |P_j(z)|$. Then, by hypothesis, $p_1 \dots p_k \leqslant 1$ and so there exists a suffix l, possibly 1 or k, such that

$$p_1 \dots p_{l-1} \geqslant p_l \dots p_k, \quad p_1 \dots p_l \leqslant p_{l+1} \dots p_k.$$

Now $P_1 \dots P_{l-1}$ and $P_l \dots P_k$ have degrees at most n, no common factor and, in view of Lemma 2, heights at most $e^n h$. Hence from Lemma 3 and the first inequality above we see that

$$p_1 \dots p_{l-1} \geqslant \exp(-4n(n+\log h)).$$

Similarly, by virtue of the second inequality above, this estimate obtains also for $p_{l+1} \dots p_k$. Thus we have

$$p_l \leqslant p_1 \dots p_k \exp(8n(n+\log h)),$$

and the assertion follows with $Q = P_l$.

Lemma 5. *If ω is a transcendental number and if $P_j(x)$ $(j = 1, 2, \dots)$ is a sequence of polynomials with degrees and heights at most n_j and h_j respectively such that*

$$n_j < n_{j+1} \ll n_j, \quad \log h_j \leqslant \log h_{j+1} \ll \log h_j,$$

then, for some infinite sequence of values of j,

$$\log |P_j(\omega)| \gg -n_j(n_j + \log h_j).$$

Here the implied constants are again absolute. For the proof we assume that the latter inequality does not hold for j sufficiently large, and we derive a contradiction if the implied constant is large enough. By Lemma 4, P_j has a factor Q_j, a power of an irreducible polynomial, such that

$$\log|Q_j(\omega)| \ll -n_j(n_j+\log h_j),$$

and, by Lemma 2, Q_j has height at most $e^{n_j}h_j$. It follows from Lemma 3 that, for all sufficiently large j, Q_j is a power of some irreducible polynomial Q, say, independent of j; for if Q_j and Q_{j+1} have no common factor then

$$\max(|Q_j(\omega)|, |Q_{j+1}(\omega)|) > e^{-4n_{j+1}^2}h_j^{-n_{j+1}}h_{j+1}^{-n_j},$$

and, in view of the hypotheses concerning n_{j+1} and h_{j+1}, this plainly contradicts either the previous inequality or its analogue with j replaced by $j+1$. Since obviously Q_j is at most the n_jth power of Q, we obtain

$$\log|Q(\omega)| \ll -(n_j+\log h_j),$$

and since also $n_j \to \infty$ as $j \to \infty$, it follows that $Q(\omega) = 0$. But this contradicts the hypothesis that ω is transcendental.

5. Main arguments

The proofs of Theorems 12.1, 12.2 and 12.3 are similar to demonstrations of earlier chapters and it will suffice therefore to describe them in outline.

For Theorem 12.1, we assume that the field generated by the ξ_i and $e^{\xi_i\eta_j}$ $(1 \leqslant i,j \leqslant 3)$ over the rationals Q has transcendence degree 1 and we derive a contradiction. The field is then generated by a transcendental number ω together with a number Ω algebraic over $Q(\omega)$; and one can assume that Ω is integral over $Q(\omega)$. It will be enough to treat here the case when the ξ_i and $e^{\xi_i\eta_j}$ are integral over $Q(\omega)$; the general result follows similarly on introducing appropriate denominators. Constants implied by $\ll$ and $\gg$, and by $c_1, c_2, \ldots$ will depend on the ξ's, η's and ω, Ω only.

One begins by constructing for any integer $k \gg 1$, an auxiliary function

$$\Phi(z) = \sum_{\lambda_0=0}^{L_0} \cdots \sum_{\lambda_3=0}^{L_3} p(\lambda_0, \ldots, \lambda_3)\,\omega^{\lambda_0} e^{(\lambda_1\xi_1+\lambda_2\xi_2+\lambda_3\xi_3)z}$$

satisfying $\Phi^{(j)}(\eta) = 0$ $(0 \leqslant j < k)$ for each

$$\eta = l_1\eta_1 + l_2\eta_2 + l_3\eta_3 \quad (1 \leqslant l_1, l_2, l_3 \leqslant m),$$

where

$$m = [k^{\frac{1}{3}}(\log k)^{\frac{1}{4}}], \quad L_0 = [k \log k], \quad L_1 = L_2 = L_3 = L = [k^{\frac{2}{3}}(\log k)^{\frac{1}{4}}],$$

and the $p(\lambda_0 \ldots, \lambda_3)$ are rational integers, not all 0, with absolute values at most $k^{c_1 k}$. Such a construction is possible, for clearly $\Phi^{(j)}(\eta)$ can be expressed as a linear form in the p's with coefficients given by polynomials in ω, Ω; the latter have degrees $\ll L_0$ in ω, $\ll 1$ in Ω and heights at most $k^{c_2 k}$. Thus one has to solve $M \ll m^3 k L_0$ linear equations in $\gg L^3 L_0 > 2M$ unknowns, and Lemma 1 of Chapter 2 is therefore applicable.

Let now C, Γ be the circles centre the origin described in the positive sense with radii k and $k^{\frac{1}{2}}$ respectively. Then, for any z on Γ,

$$\Phi(z) = \frac{1}{2\pi i}\int_C \left(\frac{A(z)}{A(\zeta)}\right)^k \frac{\Phi(\zeta)}{\zeta - z}\, d\zeta,$$

where $A(z)$ denotes the monic polynomial with m^3 zeros η. Hence we see that

$$\log |\Phi(z)| \ll -m^3 k \log k,$$

and since, by Cauchy's theorem,

$$\Phi^{(j)}(\eta) = \frac{j!}{2\pi i}\int_\Gamma \frac{\Phi(z)}{(z-\eta)^{j+1}}\, dz,$$

it follows that, if $j \leqslant k(\log k)^{\frac{1}{4}}$, then the same estimate obtains with $\Phi(z)$ replaced by $\Phi^{(j)}(\eta)$. But, by Lemma 1, Φ has $\ll L^3$ zeros within and on C, and so $\Phi^{(j)}(\eta) \neq 0$ for some η and some j as above. Further, $\Phi^{(j)}(\eta)$ is a polynomial in ω, Ω with rational integer coefficients, and, on taking the product of its conjugates over $Q(\omega)$, we derive a polynomial $P(x)$ with degree n and height h satisfying

$$n \ll k \log k, \quad \log h \ll k(\log k)^{\frac{1}{4}},$$

$$\log |P(\omega)| \ll -m^3 k \log k \ll -k^2 (\log k)^{\frac{7}{4}}.$$

As k increases we obtain a sequence of such polynomials P and, plainly, this contradicts Lemma 5. The contradiction proves the theorem.

The proof of Theorem 12.2 is similar. Under analogous initial assumptions, one constructs, for any integer $k \gg 1$, an auxiliary function

$$\Phi(z) = \sum_{\lambda_0=0}^{L_0} \cdots \sum_{\lambda_3=0}^{L_3} p(\lambda_0, \ldots, \lambda_3)\, \omega^{\lambda_0} z^{\lambda_3} e^{(\lambda_1 \xi_1 + \lambda_2 \xi_2)z}$$

satisfying $\Phi^{(j)}(\eta) = 0$ $(0 \leqslant j < k)$ for each

$$\eta = l_1\eta_1 + l_2\eta_2 \quad (1 \leqslant l_1 \leqslant m_1,\ 1 \leqslant l_2 \leqslant m_2),$$

where $$m_1 = [k^{\frac12}(\log k)^{-\frac14}], \quad m_2 = [(k\log k)^{\frac12}],$$
$$L_0 = L_3 = k, \quad L_1 = L_2 = [k^{\frac12}(\log k)^{\frac14}],$$

and the $p(\lambda_0, \ldots, \lambda_3)$ are again rational integers, not all 0, with absolute values at most $k^{c_1 k}$. The construction is certainly possible, for, in view of the hypothesis that $e^{\xi_1\eta_2}$ and $e^{\xi_2\eta_2}$ are algebraic, the coefficients in the linear forms $\Phi^{(j)}(\eta)$ have the same properties as in the previous argument, whence one has only to solve $M \ll m_1 m_2 k^2$ linear equations in $\geqslant k^3(\log k)^{\frac12} > 2M$ unknowns. Now by the first integral formula above with A denoting here the monic polynomial with $m_1 m_2$ zeros η, one has

$$\log|\Phi(z)| \ll -m_1 m_2 k \log k,$$

for all z on Γ, and, by the second integral formula, we see that the same estimate obtains with $\Phi(z)$ replaced by $\Phi^{(j)}(\eta)$ for all $j \leqslant k$ and all

$$\eta' = l_1'\eta_1 + l_2'\eta_2 \quad (1 \leqslant l_1' \leqslant m_1', \ 1 \leqslant l_2' \leqslant m_2'),$$

where $$m_1' = [k^{\frac12}(\log k)^{-\frac18}], \quad m_2' = [k^{\frac12}(\log k)^{\frac38}].$$

But, by Lemma 1, Φ has $\ll L_1 L_2 L_3$ zeros within and on C, and so $\Phi^{(j)}(\eta') \neq 0$ for some η' and some j as above. Thus, on taking conjugates over $Q(\omega)$ and appealing again to the hypothesis concerning $e^{\xi_1\eta_2}$, $e^{\xi_2\eta_2}$, we derive a polynomial $P(x)$ with degree n and height h satisfying

$$n \ll k(\log k)^{\frac12}, \quad \log h \ll k \log k,$$
$$\log|P(\omega)| \ll -m_1 m_2 k \log k \ll -k^2(\log k)^{\frac54}.$$

This contradicts Lemma 5 and the required result follows.

Finally, for the proof of Theorem 12.3, one assumes that all the $e^{\xi_i\eta_j}$ are algebraic and, adopting a notation as above, one constructs, for any integer $k \gg 1$, an auxiliary function

$$\Phi(z) = \sum_{\lambda_1=0}^{L}\sum_{\lambda_2=0}^{L}\sum_{\lambda_3=0}^{L} p(\lambda_1, \lambda_2, \lambda_3)\, e^{(\lambda_1\xi_1+\lambda_2\xi_2+\lambda_3\xi_3)z}$$

satisfying $\Phi(\eta) = 0$ for each

$$\eta = l_1\eta_1 + l_2\eta_2 \quad (1 \leqslant l_1, l_2 \leqslant k),$$

where $L = [k^{\frac34}]$, and the $p(\lambda_1, \lambda_2, \lambda_3)$ are rational integers, not all 0, with absolute values at most c_1^{Lk}. If now m is any integer $\geqslant k$ and if $\Phi(\eta) = 0$ for all η with $1 \leqslant l_1, l_2 \leqslant m$ then also $\Phi(\eta') = 0$ for all

$$\eta' = l_1'\eta_1 + l_2'\eta_2 \quad (1 \leqslant l_1', l_2' \leqslant m+1).$$

Indeed, the function Φ/A, where A denotes the monic polynomial with

m^2 zeros η, is clearly regular within and on the circle C centre the origin and radius $m^{\frac{5}{8}}$, and so, by the maximum-modulus principle or, alternatively, the first integral formula above, we have

$$\log|\Phi(\eta')| \ll -m^2 \log m;$$

on the other hand, on multiplying $\Phi(\eta')$ by a suitable denominator, one obtains an algebraic integer in a fixed field with size s satisfying $\log s \ll m^{\frac{7}{4}}$, and the assertion now follows on considering the norm of $\Phi(\eta')$. We conclude that $\Phi(\eta) = 0$ for all positive integral values of l_1, l_2, and hence $\Phi(z)$ vanishes identically. But this contradicts the hypothesis that ξ_1, ξ_2, ξ_3 are linearly independent over the rationals, and the contradiction proves the theorem.

BIBLIOGRAPHY

BOREVICH, Z. I. and SHAFAREVICH, I. R., *Number theory* (Moscow, 1964; Academic Press, London, 1966).

CASSELS, J. W. S., *An introduction to Diophantine approximation* (Cambridge Univ. Press, 1957).

CASSELS, J. W. S., *An introduction to the geometry of numbers* (Göttingen, Berlin; Springer Verlag, Heidelberg, First edn. 1959).

FELDMAN, N. I. and SHIDLOVSKY, A. B., The development and present state of the theory of transcendental numbers, *Russian Math. Surveys* **22** (1967), 1–79; translated from *Uspehi Mat. Nauk SSSR*, **22** (1967), 3–81.

GELFOND, A. O., *Transcendental and algebraic numbers* (Moscow, 1952; Dover publ., New York, 1960).

GELFOND, A. O., *Selected works* (Moscow, 1973).

GELFOND, A. O. and LINNIK, Yu. V., *Elementary methods in the analytic theory of numbers* (Moscow, 1962; Pergamon Press, Oxford, 1966).

HECKE, E., *Theorie der algebraischen Zahlen* (Teubner, Leipzig, 1923).

KOKSMA, J. F., *Diophantische Approximationen* (Ergebnisse Math., Berlin and Leipzig, 1936).

LANG, S., *Introduction to transcendental numbers* (Addison–Wesley, Reading Mass., 1966).

LANG, S., *Diophantine geometry* (Interscience, New York, 1962).

LEVEQUE, W. J., *Topics in number theory* (Addison–Wesley, Reading Mass., 1956).

MAHLER, K., *Lectures on Diophantine approximations* (Notre Dame, 1961).

MAILLET, E., *Introduction à la théorie des nombres transcendants et les propriétés arithmétiques des fonctions* (Gauthier–Villars, Paris, 1906).

NIVEN, I., *Numbers, rational and irrational* (Interscience, New York, 1961).

PERRON, O., *Die Lehre von den Kettenbrüchen* (Teubner, Leipzig, 1913).

RAMACHANDRA, K., *Lectures on transcendental numbers* (Ramanujan Inst., Univ. Madras, 1969).

SCHMIDT, W. M., *Lectures on Diophantine approximation* (Dept. Math., Univ. Colorado, 1970).

SCHNEIDER, TH., *Einführung in die transzendenten Zahlen* (Göttingen, Berlin; Springer Verlag, Heidelberg, 1957).

SIEGEL, C. L., *Transcendental numbers* (Princeton, 1949).

SKOLEM, TH., *Diophantische Gleichungen* (Ergebnisse Math., Berlin and Leipzig, 1938; reprinted: Chelsea publ., New York, 1950).

SPRINDŽUK, V. G., *Mahler's problem in metric number theory* (Minsk, 1967; American Math. Soc. Transl. Math. Monographs, vol. 25, 1969).

SPRINDŽUK, V. G. (Editor), *Current problems in the analytic theory of numbers* (Minsk, 1974).

WALDSCHMIDT, M., *Nombres transcendants*, Lecture Notes in Mathematics, vol. 402 (Springer-Verlag, Berlin, 1974).

ORIGINAL PAPERS

The following is a list of the principal papers relating to transcendental number theory. It contains most of the recent works on the subject, together with a selection of the more significant of the earlier memoirs. A fuller list of publications prior to 1967 is given in the survey article of Feldman and Shidlovsky (see Bibliography).

Abbreviations have been adopted as follows:

C.R.: *Comptes rendus de l'Académie des Sciences* (*Paris*).
D.A.N.: *Doklady Akad. Nauk SSSR.*
I.A.N.: *Izvestia Akad. Nauk SSSR.*
I.M.: *Indagationes Math.*
J.M.: *Journal reine angew. Math.*
M.A.: *Math. Annalen.*
M.Z.: *Math. Zeitschrift.*
N.A.W.: *Nederl. Akad. Wetensch. Proc. Ser. A.*
P.C.P.S.: *Proc. Cambridge Philos. Soc.*
U.M.: *Uspehi Mat. Nauk.*
V.M.: *Vestnik Moskov. Univ. Ser. I.*

I.M. is a collection of the mathematical papers in *N.A.W.*, and reference to the former is given where possible. Mathematical articles in *D.A.N.* are translated in *Soviet Math. Doklady* and, where appropriate, reference to the latter is included after an equality sign. A similar convention applies with respect to translations of articles in *Trudy Moskov.*, *Mat. Sb.*, and *Mat. Zametki*; these are available in *Transactions Moscow Math. Soc.*, *Math. USSR Sbornik*, and *Math. Notes* respectively. The abbreviation *A.M.S.T.* signifies the series *American Math. Soc. Translations* which contains some articles from other journals. Titles originally in Russian have been given in English throughout.

Adams, W. W. Transcendental numbers in the p-adic domain, *Amer. J. Math* **88** (1966), 279–308.
Simultaneous asymptotic diophantine approximations, *Mathematika*, **14** (1967), 173–80.
Asymptotic diophantine approximations and Hurwitz numbers, *Amer. J. Math.* **89** (1967), 1083–108.
Simultaneous asymptotic diophantine approximations to a basis of a real cubic number field, *J. Number Th.* **1** (1969), 179–94.
Anferteva, E. A. and Chudakov, N. G. Minima of the norm function in imaginary quadratic fields, *D.A.N.* **183** (1968), 255–6; = **9** (1968), 1342–4.
Effective estimates from below of the norms of the ideals of an imaginary quadratic field, *Mat. Sb.* **82** (1970), 55–66; = **11** (1970), 47–58.
Armitage, J. V. The Thue–Siegel–Roth theorem in characteristic p, *J. Algebra*, **9** (1968), 183–9.
An analogue of a problem of Littlewood, *Mathematika*, **16** (1969), 101–5; Corrigendum and Addendum, *ibid.* **17** (1970), 173–8.

Ax, J. On the units of an algebraic number field, *Illinois J. Math.* **9** (1965), 584–9.

On Schanuel's conjectures, *Ann. of Math.* **93** (1971), 252–68.

Babaev, G. *The distribution of integer points on algebraic surfaces*, Tadžuk. Gosudarstv. Univ. Dushanbe, 1966.

Baker, A. Continued fractions of transcendental numbers, *Mathematika*, **9** (1962), 1–8.

Rational approximations to certain algebraic numbers, *Proc. London Math. Soc.* **4** (1964), 385–98.

Rational approximations to $\sqrt[3]{2}$ and other algebraic numbers, *Quart. J. Math. Oxford*, **15** (1964), 375–83.

On Mahler's classification of transcendental numbers, *Acta Math.* **111** (1964), 97–120.

On an analogue of Littlewood's Diophantine approximation problem, *Michigan Math. J.* **11** (1964), 247–50.

Approximations to the logarithms of certain rational numbers, *Acta Arith.* **10** (1964), 315–23.

Power series representing algebraic functions, *J. London Math. Soc.* **40** (1965), 108–10.

On some Diophantine inequalities involving the exponential function, *Canadian J. Math.* **17** (1965), 616–26.

On a theorem of Sprindžuk, *Proc. Roy. Soc. London*, **A 292** (1966), 92–104.

A note on the Padé table, *I.M.* **28** (1966), 596–601.

Linear forms in the logarithms of algebraic numbers I, II, III, IV, *Mathematika*, **13** (1966), 204–16; **14** (1967), 102–7, 220–8; **15** (1968), 204–16.

On Mahler's classification of transcendental numbers: II Simultaneous Diophantine approximation, *Acta Arith.* **12** (1967), 281–8.

Simultaneous rational approximations to certain algebraic numbers, *P.C.P.S.* **63** (1967), 693–702.

A note on integral integer-valued functions of several variables, *P.C.P.S.* **63** (1967), 715–20.

Contributions to the theory of Diophantine equations: I On the representation of integers by binary forms; II The Diophantine equation $y^2 = x^3+k$, *Phil. Trans. Roy. Soc. London*, **A 263** (1968), 173–208.

The Diophantine equation $y^2 = ax^3+bx^2+cx+d$, *J. London Math. Soc.* **43** (1968), 1–9.

Bounds for the solutions of the hyperelliptic equation, *P.C.P.S.* **65** (1969), 439–44.

A remark on the class number of quadratic fields, *Bull. London Math. Soc.* **1** (1969), 98–102.

On the periods of the Weierstrass $\wp$-function, *Symposia Math.* IV, INDAM Rome, 1968 (Academic Press, London, 1970) pp. 155–74.

On the quasi-periods of the Weierstrass ζ-function, *Göttinger Nachr.* (1969) No. 16, 145–57.

An estimate for the $\wp$-function at an algebraic point, *Amer. J. Math.* **92** (1970), 619–22.

Effective methods in Diophantine problems I, II, *Proc. Symposia Pure Math.*, vol. 20 (Amer. Math. Soc., 1971), pp. 195–205; *ibid.* vol. 24, pp. 1–7.

Effective methods in the theory of numbers, *Proc. Internat. Congress Math.* Nice, 1970, vol. 1 (Gauthier-Villars, Paris, 1971), pp. 19–26.

Imaginary quadratic fields with class number 2, *Ann. Math.* **94** (1971), 139–52.

BAKER, A. On the class number of imaginary quadratic fields, *Bull. Amer. Math. Soc* **77** (1971), 678–84.

A sharpening of the bounds for linear forms in logarithms I, II, III, *Acta Arith.* **21** (1972), 117–29; **24** (1973), 33–6; **27** (1975), 247–52.

A central theorem in transcendence theory, *Diophantine approximation and its applications* (Academic Press, New York, 1973), pp. 1–23.

BAKER, A., BIRCH, B. J. and WIRSING, E. A. On a problem of Chowla, *J. Number Th.* **5** (1973), 224–36.

BAKER, A. and COATES, J. Integer points on curves of genus 1, *P.C.P.S.* **67** (1970), 595–602.

BAKER, A. and DAVENPORT, H. The equations $3x^2-2 = y^2$ and $8x^2-7 = z^2$, *Quart. J. Math. Oxford, Ser.* (2), **20** (1969), 129–37.

BAKER, A. and SCHINZEL, A. On the least integers represented by the genera of binary quadratic forms, *Acta Arith.* **18** (1971), 137–44.

BAKER, A. and SCHMIDT, W. M. Diophantine approximation and Hausdorff dimension, *Proc. London Math. Soc.* **21** (1970), 1–11.

BAKER, A. and STARK, H. M. On a fundamental inequality in number theory, *Ann. Math.* **94** (1971), 190–9.

BALKEMA, A. A. and TIJDEMAN, R. Some estimates in the theory of exponential sums, *Acta Math. Hungar.* **24** (1973), 115–133.

BARON, G. and BRAUNE, E. Zur Transzendenz von Luckenriehen mit ganzalgebraischen Koeffizienten und algebraischem Argument, *Compositio Math.* **22** (1970), 1–6, and a similar title, *Arch. Math. (Basel)* **22** (1971), 379–84.

BELOGRIVOV, I. I. Transcendence and algebraic independence of values of certain *E*-functions, *V.M.* **22** (1967), 55–62, and similar titles, *D.A.N.* **174** (1967), 267–70; = **8** (1967), 610–3; *Mat. Sb.* **82** (1970), 387–408; *Sibirsk Mat. Z.* **12** (1971), 961–82; **14** (1973), 16–35.

BOEHLE, K. Über die Transzendenz von Potenzen mit algebraischen Exponenten (Verallgemeinerung eines Satzes von A. Gelfond), *M.A.* **108** (1933), 56–74.

Über die Approximation von Potenzen mit algebraischen Exponenten durch algebraische Zahlen, *M.A.* **110** (1935), 662–78.

BOMBIERI, E. Sull'approssimazione di numeri algebrici mediante numeri algebrici, *Boll. Un. Mat. Ital.* **13** (1958), 351–4.

Algebraic values of meromorphic maps, *Invent. Math.* **10** (1970), 267–87; addendum, *ibid.* **11** (1970), 163–6.

BOMBIERI, E. and LANG, S. Analytic subgroups of group varieties, *Invent. Math.* **11** (1970), 1–14.

BOREL, E. Sur la nature arithmétique du nombre *e*, *C.R.* **128** (1899), 596 9.

BOYD, D. W. Transcendental numbers with badly distributed powers, *Proc. Amer. Math. Soc.* **23** (1969), 424–7.

BRAUER, A. Über diophantische Gleichungen mit endlich vielen Lösungen, *J.M.* **160** (1929), 70–99; **161** (1929), 1–13.

BROWNAWELL, W. Some transcendence results for the exponential function, *Norske Vid. Selsk. Skr.* (1972), no. 11.

The algebraic independence of certain values of the exponential function, *ibid.* (1972), no. 23, and a similar title, *J. Number Th.* **6** (1974), 22–31.

Sequences of Diophantine approximations, *J. Number Th.* **6** (1974), 11–21.

BRUMER, A. On the units of algebraic number fields, *Mathematika*, **14** (1967), 121–4.

BUNDSCHUH, P. Ein Satz über ganze Funktionen und Irrationalitätaussagen, *Invent. Math.* **9** (1969), 175–84.

Ein Approximationsmass für transzendente Lösungen gewisser transzendenter Gleichungen, *J.M.* **251** (1971), 32–53.

Irrationalitätsmasse für e^a, $a \neq 0$, rational oder Liouville Zahl, *M.A.* **192** (1971), 229–42.

Bundschuh, P. and Hock, A. Bestimmung aller imaginär-quadratischen Zahlkörper der Klassenzahl Eins mit Hilfe eines Satzes von Baker, *M.Z.* **111** (1969), 191–204.

Cantor, G. Über eine Eigenschaft des Inbegriffs aller reellen algebraischem Zahlen, *J.M.* **77** (1874), 258–62; = *Ges. Abh.* 116–8.

Chowla, P. Remarks on a previous paper, *J. Number Th.* **1** (1969), 522–4.

Chudakov, N. G. The upper bound for the discriminant of the tenth imaginary quadratic field with class number 1, *Studies in number theory*, no. 3. Saratov (1969), 73–7.

Chudnovsky, G. V. Algebraic independence of several values of the exponential function, *Mat. Zametki*, **15** (1974), 661–72.

Cijsouw, P. L. *Transcendence measures* (dissertation, Amsterdam, 1972).

Cijsouw, P. L. and Tijdeman, R. Distinct prime factors of consecutive integers, *Diophantine approximation and its applications* (Academic Press, London, 1973), pp. 59–76.

Transcendence of certain power series of algebraic numbers, *Acta Arith.* **23** (1973), 301–5.

Coates, J. On the algebraic approximation of functions I, II, III, IV, *I.M.* **28** (1966), 421–61; **29** (1967), 205–12.

An effective p-adic analogue of a theorem of Thue I; II: The greatest prime factor of a binary form; III: The Diophantine equation $y^2 = x^3+k$, *Acta Arith.* **15** (1969), 279–305; **16** (1970), 399–412, 425–35.

Construction of rational functions on a curve, *P.C.P.S.* **68** (1970), 105–23.

An application of the Thue–Siegel–Roth theorem to elliptic functions, *P.C.P.S.* **69** (1971), 157–61.

Linear forms in the periods of the exponential and elliptic functions, *Invent. Math.* **12** (1971), 290–9.

The transcendence of linear forms in ω_1, ω_2, η_1, η_2, $2\pi i$, *Amer. J. Math.* **93** (1971), 385–97.

Linear relations between $2\pi i$ and the periods of two elliptic curves, *Diophantine approximation and its applications* (Academic Press, London 1973), pp. 77–99.

Cugiani, M. Sulla approssimabilità dei numeri algebrici mediante numeri razionali, *Ann. Mat. Pura Appl.* **48** (1959), 135–45, and a similar title, *Boll. Un. Mat. Ital.* **14** (1959), 151–62.

Davenport, H. A note on binary cubic forms, *Mathematika*, **8** (1961), 58–62.

A note on Thue's theorem, *ibid.* **15** (1968), 76–87.

Davenport, H. and Roth, K. F. Rational approximations to algebraic numbers, *Mathematika*, **2** (1955), 160–7.

Davenport, H. and Schmidt, W. M. Approximation of real numbers by algebraic integers, *Acta Arith.* **15** (1969), 393–416.

Demjanenko, V. A. The representation of numbers by a binary cubic irreducible form, *Mat. Zametki*, **7** (1970), 87–96.

Dyson, F. J. The approximation to algebraic numbers by rationals, *Acta Math.* **79** (1947), 225–40.

Ellison, W. J., Pesek, J., Stall, D. S. and Lunnon, W. F. A postscript to a paper of A. Baker, *Bull. London Math. Soc.* **3** (1971), 75–8.

ELLISON, W. J., ELLISON, F., PESEK, J., STAHL, C. E. and STALL, D. S. The Diophantine equation $y^2+k = x^3$, *J. Number Th.* **4** (1972), 107–17.

ERDÖS, P. Representations of real numbers as sums and products of Liouville numbers, *Bull. Amer. Math. Soc.* **68** (1962), 475–8.

ERDÖS, P., and MAHLER, K. Some arithmetical properties of the convergents of a continued fraction, *J. London Math. Soc.* **14** (1939), 12–18.

FADDEEV, D. K. On a paper by A. Baker, *Zap. Naučn. Sem. Leningrad Otdel. Mat. Inst. Steklov*, **1** (1966), 128–39.

FELDMAN, N. I. The approximation of some transcendental numbers I, II, *I.A.N.* **15** (1951), 53–74; = *A.M.S.T.* **59** (1969), 224–70.

On the simultaneous approximation of the periods of elliptic functions by algebraic numbers, *I.A.N.* **22** (1958), 563–76; = *A.M.S.T.* **59** (1966), 271–84.

On the measure of transcendence of π, *I.A.N.* **24** (1960), 357–68; = *A.M.S.T.* **58** (1966), 110–24.

On the approximation by algebraic numbers of the logarithms of algebraic numbers, *I.A.N.* **24** (1960), 475–92; = *A.M.S.T.* **58** (1966), 125–42.

On transcendental numbers having approximations of a given type, *U.M.* **17** (1962), 145–51.

On the problem of the measure of transcendence of e, *U.M.* **18** (1963), 207–13.

Arithmetic properties of the solutions of some transcendental equations, *V.M.* **1** (1964), 13–20.

An estimate of the absolute value of a linear form in the logarithms of certain algebraic numbers, *Mat. Zametki*, **2** (1967), 245–56, and a similar title, *V.M.* **2** (1967), 63–72.

Estimation of a linear form in the logarithms of algebraic numbers, *Mat. Sb.* **76** (1968), 304–19; = **5** (1968), 291–307, and similar titles, *Mat. Sb.* **77** (1968), 423–36; = **6** (1968), 393–406; *U.M.* **23** (1968), 185–6; *D.A.N.* **182** (1968), 1278–9; = **9** (1968), 1284–5.

An elliptic analogue of an inequality of A. O. Gelfond, *Trudy Moskov.* **18** (1968), 65–76; = **18** (1968), 71–83.

Refinement of two effective inequalities of A. Baker, *Mat. Zametki*, **6** (1969), 767–9, and related work, **5** (1969), 681–9.

Effective bounds for the size of the solutions of certain Diophantine equations, *Mat. Zametki*, **8** (1970), 361–71; = **8** (1970), 674–9, and related work, *ibid.* **7** (1970), 569–80; = **7** (1970), 343–9; *V.M.* **26** (1971), 52–8.

An effective refinement of the exponent in Liouville's theorem, *I.A.N.* **35** (1971), 973–90; = *Math. USSR Izv.* **5** (1971), 985–1002; and related work *D.A.N.* **207** (1972), 41–3.

On a certain linear form, *Acta Arith.* **21** (1972), 347–55.

FELDMAN, N. I. and CHUDAKOV, N.G. On a theorem of Stark, *Mat. Zametki*, **3** (1972), 329–40.

FRAENKEL, A. S. On a theorem of D. Ridout in the theory of Diophantine approximations, *Trans. Amer. Math. Soc.* **105** (1962), 84–101.

Transcendental numbers and a conjecture of Erdös and Mahler, *J. London Math. Soc.* **39** (1964), 405–16.

Distance to the nearest integer and algebraic independence of certain real numbers, *Proc. Amer. Math. Soc.* **16** (1965), 154–60.

FRAENKEL, A. S. and BOROSH, I. Fractional dimension of a set of transcendental numbers, *Proc. London Math. Soc.* **15** (1965), 458–70.

GALOČKIN, A. I. Bounds for the measure of transcendence of the values of E-functions, *Mat. Zametki*, **3** (1968), 377–86, and similar titles, *ibid.* **8** (1970), 19–28; = **8** (1970), 478–84; *V.M.* **25** (1970), 58–63.

Gelfond, A. O. Sur les propriétés arithmetiques des fonctions entières, *Tôhoku Math. J.* **30** (1929), 280–5.

Sur les nombres transcendants, *C.R.* **189** (1929), 1224–6.

Sur le septième problème de Hilbert, *I.A.N.* **7** (1934), 623–30; *D.A.N.* **2** (1934), 1–6.

On the approximation of transcendental numbers by algebraic numbers, *D.A.N.* **2** (1935), 177–82.

On the approximation by algebraic numbers of the ratio of the logarithms of two algebraic numbers, *I.A.N.* **5–6** (1939), 509–18.

Sur la divisibilité de la différence des puissances de deux nombres entières par une puissance d'un idéal premier, *Mat. Sb.* **7** (1940), 7–26.

The approximation of algebraic numbers by algebraic numbers and the theory of transcendental numbers, *U.M.* **4** (1949), 19–49; = *A.M.S.T.* **2** (1962), 81–124.

On the algebraic independence of algebraic powers of algebraic numbers, *D.A.N.* **64** (1949), 277–80, and similar titles, *D.A.N.* **67** (1949), 13–14; *U.M.* **5** (1949), 14–48; = *A.M.S.T.* **2** (1962), 125–69.

Gelfond, A. O. and Feldman, N. I. On lower bounds for linear forms in three logarithms of algebraic numbers, *V.M.* **5** (1949), 13–16.

On the measure of mutual transcendence of certain numbers, *I.A.N.* **14** (1950), 493–500.

Gelfond, A. O. and Linnik, Yu. V. On Thue's method and the problem of effectiveness in quadratic fields, *D.A.N.* **61** (1948), 773–6.

Goldstein, L. J. A generalization of Stark's theorem, *J. Number Th.* **3** (1971), 323–46.

Imaginary quadratic fields of class number 2, *ibid.* **4** (1972), 286–301.

Gordan, P. Transzendenz von e und π, *M.A.* **43** (1893), 222–5.

Günther, A. Über transzendente $\mathfrak{p}$-adische Zahlen I, II, *J.M.* **192** (1953), 155–66; **193** (1954), 1–10.

Güting, R. Approximation of algebraic numbers by algebraic numbers, *Michigan Math. J.* **8** (1961), 149–59.

On Mahler's function θ_1, *ibid.* **10** (1963), 161–79.

Über die Zusammenhang zwischen rationalen Approximationen und Kettenbrüchentwicklungen, *M.Z.* **90** (1965), 382–7.

Polynomials with multiple zeros, *Mathematika*, **14** (1967), 181–96.

Hanecke, W. Über ein System simultaner diophantischer Approximationen, *M.Z.* **110** (1969), 378–84.

Hermite, Ch. Sur la fonction exponentielle, *C.R.* **77** (1873), 18–24, 74–9, 226–33, 285–93; = *Oeuvres* III, 150–81.

Hilbert, D. Über die Transzendenz der Zahlen e und π, *Göttinger Nachr.* (1893), 113–16; *M.A.* **43** (1893), 216–20; = *Ges. Abh.* I, 1–4.

Hilliker, D. L. On analytic functions which have algebraic values at a convergent sequence of points, *Trans. Amer. Math. Soc.* **126** (1967), 534–50.

Hurwitz, A. Ueber arithmetische Eigenschaften gewisser transcendenter Functionen I, II, *M.A.* **22** (1883), 211–29; **32** (1888), 583–8.

Beweis der Transzendenz der Zahl e, *Göttinger Nachr.* (1893), 153–5.

Hyyrö, S. Über Approximation algebraischer Zahlen durch rationale, *Ann. Univ. Turku.* ser. AI, **84** (1965), 1–12, and similar titles, *Ann. Acad. Sci. Fenn.* ser. AI, nos. 376 (1965), 394 (1967).

Içen, O. S. Eine Verallgemeinerung und Übertragung der Schneiderschen Algebraizitätskriterien ins p-adische mit Anwendung auf einen Transzendenzbeweis im p-adischen, *J.M.* **198** (1957), 28–55; Berichtigung, *ibid.* 220.

Kappe, Luise-Charlotte. Zur Approximation von e^a, *Ann. Univ. Sci. Budapest Eötvös Sect. Math.* **9** (1966), 3–14.

Kasch, F. Zur Annäherung algebraischer Zahlen durch arithmetisch charakterisierte rationale Zahlen, *Math. Nachr.* **10** (1953), 85–98.

Über eine metrische Eigenschaft der S-Zahlen, *M.Z.* **70** (1958), 263–70, and a similar title, *J.M.* **203** (1960), 157–9.

Kasch, F. and Volkmann, B. Zur Mahlerschen Vermutung über S-Zahlen, *M.A.* **136** (1958), 442–53.

Kaufman, R. M. Bounds for linear forms in the logarithms of algebraic numbers with p-adic metric, *V.M.* **26** (1971), 3–10.

Keates, M. On the greatest prime factor of a polynomial, *Proc. Edinburgh Math. Soc.* **16** (1969), 301–3.

Koksma, J. Über die Mahlersche Klasseneinteilung der transzendenten Zahlen und die Approximation komplexer Zahlen durch algebraische Zahlen, *Monatsh. Math. Phys.* **48** (1939), 176–89.

Koksma, J. F. and Popken, J. Zur Transzendenz von e^π, *J.M.* **168** (1932), 211–30.

Kuzmin, R. O. On a new class of transcendental numbers, *I.A.N.* **3** (1930), 583–97.

Lambert, J. H. Mémoire sur quelques propriétés remarquables des quantités transcendantes circulaires et logarithmiques, *Histoire Acad. roy. sci. et belles lettr.* Berlin, Année 1761 (1768), 265–322.

Lang, S. On a theorem of Mahler, *Mathematika*, **7** (1960), 139–40.

A transcendence measure for E-functions, *ibid.* **9** (1962), 157–61.

Transcendental points on group varieties, *Topology*, **1** (1962), 313–8.

Asymptotic approximations to quadratic irrationalities I, II, *Amer. J. Math.* **87** (1965), 481–95.

Report on Diophantine approximations, *Bull. Soc. Math. France*, **93** (1965), 177–92.

Algebraic values of meromorphic maps I, II, *Topology*, **3** (1965), 183–91; **5** (1966), 363–70.

Transcendental numbers and Diophantine approximations, *Bull. Amer. Math. Soc.* **77** (1971), 635–77.

LeVeque, W. J. Note on the transcendence of certain series, *Proc. Amer. Math. Soc.* **2** (1951), 401–3.

Note on S-numbers, *Proc. Amer. Math. Soc.* **4** (1953), 189–90.

On Mahler's U-numbers, *J. London Math. Soc.* **28** (1953), 220–9.

Levin, B. V., Feldman, N. I. and Shidlovsky, A. B. Alexander O. Gelfond, *Acta Arith.* **17** (1971), 315–36.

Lindemann, F. Über die Zahl π, *M.A.* **20** (1882), 213–25.

Über die Ludolph'sche Zahl, *S.-B. Preuss Akad. Wiss.* (1882), 679–82.

Linnik, Yu. V. and Chudakov, N. G. On a class of completely multiplicative functions, *D.A.N.* **74** (1950), 193–6.

Liouville, J. Sur l'irrationalité du nombre e, *J. Math. pures appl.* **5** (1840), 192–3.

Sur des classes très-étendues de quantités dont la valeur n'est ni algébrique, ni même reductible à des irrationnelles algébriques, *C.R.* **18** (1844), 883–5, 910–11, *J. Math. pures appl.* **16** (1851), 133–42.

Mahler, K. Über Beziehungen zwischen der Zahl e und den Liouvilleschen Zahlen, *M.Z.* **31** (1930), 729–32.

Ein Beweis des Thue–Siegelschen Satzes über die Approximation algebraischen Zahlen für binomische Gleichungen, *M.A.* **105** (1931), 267–76.

Zur Approximation der Exponentialfunktion und des Logarithmus I, II, *J.M.* **166** (1932), 118–50.

Ein Beweis der Transzendenz der p-adischen Exponentialfunktion, *J.M.* **169** (1932), 61–6.

Über das Mass der Menge aller S-Zahlen, *M.A.* **106** (1932), 131–9.

Zur Approximation algebraischer Zahlen: I Über den grössten Primteiler binärer Formen, II Über die Anzahl der Darstellungen grosser Zahlen durch binäre Formen, III Über die mittlere Anzahl der Darstellungen grosser Zahlen durch binäre Formen, *M.A.* **107** (1933), 691–730; **108** (1933), 37–55; *Acta Math.* **62** (1933), 91–166.

Über transzendente p-adische Zahlen, *Compositio Math.* **2** (1935), 259–75, and related work, *Mathematica (Leiden)*, **3** (1935), 177–85.

Ein Analogon zu einen Schneiderschen Satz, *N.A.W.* **39** (1936), 633–40, 729–37.

Arithmetische Eigenschaften einer Klasse von Dezimalbrüchen, *N.A.W.* **40** (1937), 421–8, and related work, *Math. B. Zutphen*, **6** (1937), 22–36.

On the fractional parts of the powers of a rational number I, II, *Acta Arith.* **3** (1938), 89–93; *Mathematika*, **4** (1957), 122–4.

On the solution of algebraic differential equations, *N.A.W.* **42** (1939), 61–3.

On the theorem of Liouville in fields of positive characteristic, *Canadian J. Math.* **1** (1949), 397–400.

On Dyson's improvement of the Thue–Siegel theorem, *I.M.* **11** (1949), 449–58.

On the continued fractions of quadratic and cubic irrationals, *Ann. Math. pura appl.* **30** (1949), 147–72.

On the generating functions of the integers with a missing digit, *J. Indian Math. Soc.* **A 15** (1951), 33–40.

On the greatest prime factor of ax^m+by^n, *N.A.W.* **1** (1953), 113–22.

On the approximation of logarithms of algebraic numbers, *Phil. Trans. Roy. Soc. London*, **A 245** (1953), 371–98.

On the approximation of π, *I.M.* **15** (1953), 30–42.

On a theorem of E. Bombieri, *I.M.* **22** (1960), 245–53; Correction, *ibid.* **23** (1961), 141.

An application of Jensen's formula to polynomials, *Mathematika*, **7** (1960), 98–100.

On some inequalities for polynomials in several variables, *J. London Math. Soc.* **37** (1962), 341–4.

On the approximation of algebraic numbers by algebraic integers, *J. Australian Math. Soc.* **1** (1963), 408–34.

Transcendental numbers, *ibid.* **4** (1964), 393–6.

An inequality for a pair of polynomials that are relatively prime, *ibid.* **4** (1964), 418–20.

Arithmetic properties of lacunary power series with integral coefficients, *ibid.* **5** (1965), 56–64.

Applications of some formulae by Hermite to the approximation of exponentials and logarithms, *M.A.* **168** (1967), 200–27.

On a class of entire functions, *Acta Math. Hungar.* **18** (1967), 83–96.

A remark on recursive sequences, *J. Math. Sci.* **1** (1966), 12–17.

Applications of a theorem of A. B. Shidlovsky, *Proc. Roy. Soc. London*, **A 305** (1968), 149–73, and a similar title, *Mat. Zametki*, **2** (1967), 25–32.

Perfect systems, *Compositio Math.* **19** (1968), 95–166.

Remarks on a paper by W. Schwarz, *J. Number Th.* **1** (1969), 512–21.

MAHLER, K. On the order function of a transcendental number, *Acta Arith.* **18** (1971), 63–76.

MAHLER, K. and POPKEN, J. Ein neues Prinzip für Transzendenzbeweise *N.A.W.* **38** (1935), 864–71.

MAHLER, K. and SZEKERES, G. On the approximation of real numbers by roots of integers, *Acta Arith.* **12** (1967), 315–20.

MAIER, W. Potenzreihen irrationalen Grenzwertes, *J.M.* **156** (1927), 93–148.

MAILLET, E. Sur les racines des équations transcendantes à coefficients rationnels, *J. Math. pures appl.* **7** (1901), 419–40; *C.R.* **132** (1901), 908–10; **133** (1901), 989–90.

Sur les propriétés arithmétiques des fonctions entières et quasi entières, *C.R.* **134** (1902), 1131–2; *Bull. Soc. Math. France,* **30** (1902), 134–53.

Sur les fonctions monodromes et les nombres transcendants, *J. Math. pures appl.* **10** (1904), 275–362; *C.R.* **138** (1904), 262–5.

Sur les nombres e et π, et les equations transcendants, *C.R.* **133** (1901), 1191–2; *Acta Math.* **29** (1905), 295–331.

Sur la classification des irrationnelles, *C.R.* **143** (1906), 26–8.

Sur les nombres transcendants dont le développement en fraction continue est quasi-périodique, et sur les nombres de Liouville, *Bull. Soc. Math. France,* **34** (1906), 213–27, and related work *ibid.* **35** (1907), 27–47; **50** (1922), 74–99; *J. Math. pures appl.* **3** (1907), 299–336; *C.R.* **170** (1920), 983–6.

MASSER, D. W. On the periods of the exponential and elliptic functions, *P.C.P.S.* **73** (1973), 339–50.

Elliptic functions and transcendence, Ph.D. dissertation (Cambridge, 1974).

MENDÈS FRANCE, M. Nombres transcendants et ensembles normaux, *Acta Arith.* **15** (1969), 189–92.

MEYER, Y. Nombres algébriques, nombres transcendants et equirépartition modulo 1, *Acta Arith.* **16** (1970), 347–50.

MORDUCHAI-BOLTOVSKOJ, D. Sur le logarithme d'un nombre algébrique, *C.R.* **176** (1923), 724–7.

On some properties of transcendental numbers of the second class, *Mat. Sb.* **34** (1927), 55–100.

On transcendental numbers with successive approximations defined by algebraic equations, *ibid.* **41** (1934), 221–32.

Über einige Eigenschaften der transzendenten Zahlen, *Tôhoku Math. J.* **40** (1935), 99–127.

On conditions for the definition of a number by transcendental equations of some general type *D.A.N.* **52** (1946), 487–90.

On hypertranscendental functions and hypertranscendental numbers, *D.A.N.* **64** (1949), 21–4.

MORITA, Y. On transcendency of special values of arithmetic automorphic functions, *J. Math. Soc. Japan* **24** (1972), 268–74.

NESTERENKO, JU. V. The algebraic independence of the values of E-functions which satisfy linear homogeneous differential equations, *Mat. Zametki* **5** (1969), 587–98.

VON NEUMANN, J. Ein System algebraisch unabhängiger Zahlen, *M.A.* **99** (1928), 134–41.

NURMAGOMEDOV, M. S. The arithmetic properties of a certain class of analytic functions, *Mat. Sb.* **85** (1971), 339–65, and similar titles, *V.M.* **26**, no. 6 (1971), 79–86; **28**, no. 1 (1973), 19–25.

OLEINIKOV, V. A. On some properties of algebraically dependent quantities, *V.M.* **5** (1962), 11–17.

On the transcendence and algebraic independence of the values of some *E*-functions, *V.M.* **6** (1962), 34–8.

On the algebraic independence of the values of *E*-functions that are solutions of third-order linear inhomogeneous differential equations, *D.A.N.* **169** (1966), 32–4; = **7** (1966), 869–71, and related work, *ibid.* **166** (1966), 540–3; *Mat. Sb.* **78** (1969), 301–6.

OSGOOD, C. F. Some theorems on diophantine approximation, *Trans. Amer. Math. Soc.* **123** (1966), 64–87.

A method in diophantine approximation I–V, *Acta Arith.* **12** (1967), 111–29; **13** (1968), 383–93; **16** (1970), 5–40; **22** (1973), 353–69.

The simultaneous diophantine approximation of certain *k*th roots, *P.C.P.S.* **67** (1970), 75–86.

On the diophantine approximation of values of functions satisfying certain *q*-difference equations, *J. Number Th.* **3** (1971), 159–77.

On the simultaneous diophantine approximation of values of certain algebraic functions, *Acta Arith.* **19** (1971), 343–86.

A Diophantine property of $J_0(z)$, *Diophantine approximation and its applications* (Academic Press, London, 1973), pp. 201–9.

An effective lower bound on the 'Diophantine approximation' of algebraic functions by rational functions, *Mathematika*, **20** (1973), 4–15.

PARRY, C. J. The *p*-adic generalisation of the Thue–Siegel theorem, *J. London Math. Soc.* **15** (1940), 293–305, and a similar title, *Acta Math.* **83** (1950), 1–100.

PECK, L. G. Simultaneous rational approximations to algebraic numbers, *Bull. Amer. Math. Soc.* **67** (1961), 197–201.

PERSHIKOVA, T. V. On the transcendence of values of certain *E*-functions, *V.M.* **21** (1966), 55–61.

PÓLYA, G. Über ganzwertige ganze Funktionen, *Rend. Circ. Mat. Palermo*, **40** (1915), 1–16.

VAN DER POORTEN, A. J. Transcendental entire functions mapping every algebraic number field into itself, *J. Australian Math. Soc.* **8** (1968), 192–3.

On the arithmetic nature of definite integrals of rational functions, *Proc. Amer. Math. Soc.* **29** (1971), 451–6.

POPKEN, J. Sur la nature arithmétique du nombre *e*, *C.R.* **186** (1928), 1505–7.

Zur Transzendenz von *e*, *M.Z.* **29** (1929), 525–41.

Zur Transzendenz von π, *M.Z.* **29** (1929), 542–8.

Eine arithmetische Eigenschaft gewisser ganzer Funktionen I, II, *N.A.W.* **40** (1937), 142–50, 263–70.

On Lambert's proof for the irrationality of π, *N.A.W.* **43** (1940), 712–14; **52** (1949), 504.

An arithmetical theorem concerning linear differential equations of infinite order, *I.M.* **12** (1950), 1645–6.

Un théorème sur les nombres transcendants, *Bull. Soc. Math. Belg.* **7** (1955), 124–30.

Arithmetical properties of the Taylor coefficients of algebraic functions, *I.M.* **21** (1959), 202–10.

Irrational power series, *I.M.* **25** (1963), 691–4.

Note on a generalization of a problem of Hilbert, *I.M.* **27** (1965), 178–81.

Algebraic independence of certain zêta functions, *I.M.* **28** (1966), 1–5.

A measure for the differential transcendence of the zeta-function of Riemann,

Number theory and analysis (Papers in honour of Edmund Landau), (Plenum, New York, 1969), pp. 245–55.

Popken, J. A contribution to the Thue–Siegel–Roth problem, *Number theory* (North-Holland, Amsterdam, 1970), pp. 181–90.

Ramachandra, K. Approximation of algebraic numbers, *Göttinger Nachr.* (1966), 45–52.

Contributions to the theory of transcendental numbers I, II, *Acta Arith.* **14** (1968), 65–88.

A note on Baker's method, *J. Australian Math. Soc.* **10** (1969), 197–203.

A lattice point problem for norm forms in several variables, *J. Number Th.* **1** (1969), 534–55.

A note on numbers with a large prime factor I, II, III, *J. London Math. Soc.* **1** (1969), 303–6; *J. Indian Math. Soc.* **34** (1970), 39–48; *Acta Arith.* **19** (1971), 49–62.

Application of Baker's theory to two problems considered by Erdös and Selfridge, *J. Indian Math. Soc.* **37** (1973) 25–34.

Ramachandra, K. and Shorey, T. N. On gaps between numbers with a large prime factor, *Acta Arith.* **24** (1973), 99–111.

Rauzy, G. Algébricité des fonctions méromorphes prenant certaines valeurs algébriques, *Bull. Soc. Math. France*, **96** (1968), 197–208.

Ridout, D. Rational approximations to algebraic numbers, *Mathematika*, **4** (1957), 125–31.

The p-adic generalization of the Thue–Siegel–Roth theorem, *ibid.* **5** (1958), 40–8.

Roth, K. F. Rational approximations to algebraic numbers, *Mathematika*, **2** (1955), 1–20; corrigendum, *ibid.* 168.

Schinzel, A. On two theorems of Gelfond and some of their applications, *Acta Arith.* **13** (1968), 177–236.

An improvement on Runge's theorem on Diophantine equations, *Comment. Pontificia Acad. Sci.* **2** (1969), no. 20, 1–9.

Primitive divisors of the expression $A^n - B^n$ in algebraic number fields, *J.M.* **269** (1974), 27–33.

Schmidt, W. M. Simultaneous approximation and algebraic independence of numbers, *Bull. Amer. Math. Soc.* **68** (1962), 475–8.

Über simultane Approximation algebraischer Zahlen durch rationale, *Acta Math.* **114** (1965), 159–206.

Simultaneous approximation to a basis of a real number field, *Amer. J. Math.* **88** (1966), 517–27.

Some diophantine equations in three variables with only finitely many solutions, *Mathematika*, **14** (1967), 113–20.

T-numbers do exist, *Symposia Math.* IV, INDAM, Rome, 1968 (Academic Press, London, 1970), pp. 3–26.

Simultaneous approximation to algebraic numbers by rationals, *Acta Math.* **125** (1970), 189–201.

Mahler's T-numbers, *Proc. Symposia Pure Math.*, vol. 20 (Amer. Math. Soc. 1971), pp. 275–86.

Linearformen mit algebraischen Koeffizienten I, II, *J. Number Th.* **3** (1971), 253–77; *M.A.* **191** (1971), 1–20.

Norm form equations, *Ann. Math.* **96** (1972), 526–51.

Schneider, Th. Transzendenzuntersuchungen periodischer Funktionen: I Transzendenz von Potenzen; II Transzendenzeigenschaften elliptischer Funktionen, *J.M.* **172** (1934), 65–74.

Über die Approximation algebraischer Zahlen, *J.M.* **175** (1936), 182–92.
Arithmetische Untersuchungen elliptischer Integrale, *M.A.* **113** (1937), 1–13.
Zur Theorie der Abelschen Funktionen und Integrale, *J.M.* **183** (1941), 110–28.
Über eine Dysonsche Verschärfung des Siegel–Thueschen Satzes, *Arch. Math.* **1** (1949), 288–95.
Ein Satz über ganzwertige Funktionen als Prinzip für Transzendenzbeweise, *M.A.* **121** (1949), 131–40.
Zur Annäherung der algebraischen Zahlen durch rationale, *J.M.* **188** (1950), 115–28.
Anwendung eines abgeänderten Roth–Ridoutschen Satzes auf diophantische Gleichungen, *M.A.* **169** (1967), 177–82.
Über p-adische Kettenbrüche, *Symposia Math.* IV, INDAM, Rome, 1968 (Academic Press, London, 1970), pp. 181–9.

SCHWARZ, W. Remarks on the irrationality and transcendence of certain series, *Math. Scand.* **20** (1967), 269–74.

SHIDLOVSKY, A. B. On bounds for the measure of transcendence of a subclass of the numbers α^β, *V.M.* **6** (1951), 17–28.
On the transcendence and algebraic independence of the values of integral functions of some classes, and similar titles, *D.A.N.* **96** (1954), 697–700; **100** (1955), 221–4; **103** (1955), 979–80; **105** (1955), 35–7; **108** (1956), 400–3.
On a criterion for algebraic independence of the values of a class of integral functions, *I.A.N.* **23** (1959), 35–66; = *A.M.S.T.* **22** (1962), 339–70.
On the transcendence and algebraic independence of some E-functions, and similar titles, *V.M.* **5** (1960), 19–28; **5** (1961), 44–59; *I.A.N.* **26** (1962), 877–910; = *A.M.S.T.* **50** (1966), 141–77.
On a generalization of Lindemann's theorem, *D.A.N.* **138** (1961), 1301–4; = **2** (1961), 841–4.
A general theorem on the algebraic independence of E-functions, and similar titles, *D.A.N.* **169** (1966), 42–5; **171** (1966), 810–13; *Litovsk Mat. Sb.* **1** (1966), 129–30.
On estimates of the measure of transcendence of E-functions, *U.M.* **22** (1967), 245–56; *Mat. Zametki*, **2** (1967), 33–44.
Algebraic independence of the values of certain hypergeometric E-functions, *Trudy Moskov*, **18** (1968), 55–64; = **18** (1968), 59–69.
Transcendence and algebraic independence of values of E-functions, *Proc. Internat. Congress Math.* (Moscow, 1968), pp. 299–307.
A certain theorem of C. Siegel, *V.M.* **24** (1969), 39–42.

SHOREY, T. N. On a theorem of Ramachandra, *Acta Arith.* **20** (1972), 215–21.
Linear forms in the logarithms of algebraic numbers with small coefficients I, II, *J. Indian Math. Soc.* **38** (1974), 271–92.
Algebraic independence of certain numbers in the p-adic domain, *I.M.* **34** (1972), 423–35, and related work, *ibid.* 436–42.
On gaps between numbers with a large prime factor II, *Acta Arith.* **25** (1974), 365–73.

SIEGEL, C. L. Approximation algebraischer Zahlen, *M.Z.* **10** (1921), 173–213.
Über Näherungswerte algebraischer Zahlen, *M.A.* **84** (1921), 80–99.
Über den Thueschen Satz, *Norske Vid. Selsk. Skr.*, no. 16 (1921).
Über einige Anwendungen diophantischer Approximationen, *Abh. Preuss. Akad. Wiss.* no. 1 (1929).
Über die Perioden elliptischer Funktionen, *J.M.* **167** (1932), 62–9.

Siegel, C. L. The integer solutions of the equation $y^2 = ax^n + bx^{n-1} + \ldots + k$, *J. London Math. Soc.* **1** (1926), 66–8 (under the pseudonym X).

Die Gleichung $ax^n - by^n = c$, *M.A.* **114** (1937), 57–68.

Abschatzung von Einheiten, *Göttinger Nachr.* (1969), 71–86.

Einige Erläuterungen zu Thues Untersuchungen über Annäherungswerte algebraischer Zahlen und diophantische Gleichungen, *ibid.* (1970), 169–95.

Slesoraitene, R. The Mahler–Sprindžuk theorem for polynomials of the third degree in two variables, and similar titles, *Litovsk. Mat. Sb.* **9** (1969), 627–34; **10** (1970), 367–74, 545–64, 791–813; **13** (1973), 177–88.

Šmelev, A. A. The algebraic independence of a certain class of transcendental numbers, and similar titles, *Mat. Zametki*, **3** (1968), 51–8; **4** (1968), 341–8, 525–32; **5** (1969), 117–28; **7** (1970), 203–10.

Algebraic independence of values of certain *E*-functions, *Izv. Vysš Učebn. Zaved. Matematika* (1969), no. 4 (83), 103–12.

Spira, R. A lemma in transcendental number theory, *Trans. Amer. Math. Soc.* **146** (1949), 457–64 (but see review no. 5307 in *Math. Reviews*, 41).

Sprindžuk, V. G. On some general problems of approximating numbers by algebraic numbers, *Litovsk Mat. Sb.* **2** (1962), 129–45.

On a classification of transcendental numbers, *ibid.* **2** (1962), 215–9.

A proof of Mahler's conjecture on the measure of the set of *S*-numbers, and similar titles, *ibid.* **2** (1962), 221–6; *D.A.N.* **154** (1964), 783–6; **155** (1964), 54–6; = **5** (1964), 183–7, 361–3; *U.M.* **19** (1964), 191–4; *I.A.N.* **29** (1965), 379–436; = *A.M.S.T.* **51** (1966), 215–72.

A metric theorem on the smallest integral polynomials of several variables, *Dokl. Akad. Nauk BSSR*, **11** (1967), 5–6.

Concerning Baker's theorem on linear forms in logarithms, *ibid.* **11** (1967), 767–9.

The finiteness of the number of rational and algebraic points on certain transcendental curves, *D.A.N.* **177** (1967), 524–7.

Estimates of linear forms with *p*-adic logarithms of algebraic numbers, *Vesci Akad. Navuk BSSR*, Ser. Fiz-Mat. (1968), no. 4, 5–14.

Irrationality of the values of certain transcendental functions, *I.A.N.* **32** (1968), 93–107.

Effectivization in certain problems of Diophantine approximation theory, *Dokl. Akad. Nauk BSSR*, **12** (1968), 293–7.

On the metric theory of 'nonlinear' diophantine approximations, *ibid.* **13** (1969), 298–301.

On the theory of the hypergeometric functions of Siegel, *ibid.* **13** (1969), 389–91.

Effective estimates in 'ternary' exponential Diophantine equations, *ibid.* **13** (1969), 777–80.

Effective bounds on rational approximations to algebraic numbers, *ibid.* **14** (1970), 681–4, and similar titles, *ibid.* **15** (1971), 101–4; *I.A.N.* **35** (1971), 991–1007.

A new application of *p*-adic analysis to representations of numbers by binary forms, *I.A.N.* **34** (1970), 1038–63.

New applications of analytic and *p*-adic methods in Diophantine approximations, *Proc. Internat. Congress Math.* Nice, 1970, vol. 1 (Gauthier-Villars, Paris, 1971).

On the largest prime factor of a binary form, *Dokl. Akad. Nauk BSSR*, **15** (1971), 389–91, and related work **17** (1973), 685–8.

On bounds for the units in algebraic number fields, *ibid.* **15** (1971), 1065–8.

On bounds for the solutions of the Thue equation, *I.A.N.* **36** (1972), 712–41.
The method of trigonometrical sums in the metrical theory of Diophantine approximation, *Trudy Mat. Inst. Steklov*, **128** (1972), 212–28.
STARK, H. M. A complete determination of the complex quadratic fields with class-number one, *Michigan Math. J.* **14** (1967), 1–27.
On the 'gap' in a theorem of Heegner, *J. Number Th.* **1** (1969), 16–27.
A historical note on complex quadratic fields with class-number one, *Proc. Amer. Math. Soc.* **21** (1969), 254–5.
A transcendence theorem for class number problems I, II, *Ann. Math.* **94** (1971), 153–73; **96** (1972), 174–209.
Effective estimates of solutions of some diophantine equations, *Acta Arith.* **24** (1973), 251–9.
Further advances in the theory of linear forms in logarithms, *Diophantine approximation and its applications* (Academic Press, London, 1973), pp. 255–93.
STEPANOV, S. A. The approximation of an algebraic number by algebraic numbers of a special form, *V.M.* **22** (1967), 78–86.
STRAUS, E. G. On entire functions with algebraic derivatives at certain algebraic points, *Ann. Math.* **52** (1950), 188–98.
TARTAKOVSKY, V. A. A uniform estimate for the number of representations of unity by a binary form of degree $n \geqslant 3$, *D.A.N.* **193** (1970), 764; = **11** (1970), 1026–7.
THUE, A. Bemerkungen über gewisse Näherungsbrüche algebraischer Zahlen, *Norske Vid. Selsk. Skr.* No. 3 (1908).
Über rationale Annäherungswerte der reellen Wurzel der ganzen Funktion dritten Grades $x^3 - ax - b$, *ibid.* no. 6 (1908) (see also no. 7).
Über Annäherungswerte algebraischer Zahlen, *J.M.* **135** (1909), 284–305.
Ein Fundamentaltheorem zur Bestimmung von Annäherungswerten aller Wurzeln gewisser ganzer Funktionen, *J.M.* **138** (1910), 96–108.
Über eine Eigenschaft, die kiene transzendente Grösse haben kann, *Norske Vid. Selsk. Skr.* no. 20 (1912).
Berechnung aller Lösungen gewisser Gleichungen von der Form $ax^r - by^r = f$, *ibid.* no. 4 (1918).
TIJDEMAN, R. On the number of zeros of general exponential polynomials, *I.M.* **33** (1971), 1–7.
On the algebraic independence of certain numbers, *I.M.* **33** (1971), 146–62.
On the maximal distance of numbers with a large prime factor, *J. London Math. Soc.* **5** (1972), 313–20.
An auxiliary result in the theory of transcendental numbers, *J. Number Th.* **5** (1973), 80–94.
On integers with many small prime factors, *Compositio Math.* **26** (1973), 319–30.
UCHIYAMA, S. On the Thue–Siegel–Roth theorem I, II, III, *Proc. Japan Acad.* **35** (1959), 413–6, 525–9; **36** (1960), 1–2.
VELDKAMP, G. R. Ein Transzendenz-Satz für p-adische Zahlen, *J. London Math. Soc.* **15** (1940), 183–92.
VINOGRADOV, A. I. and SPRINDŽUK, V. G. The representation of numbers by binary forms, *Mat. Zametki*, **3** (1968), 369–76.
VOLKMANN, B. Zur kubischen Fall der Mahlerschen Vermutung, *M.A.* **139** (1959), 87–90.
Zur Mahlerschen Vermutung im Komplèxen, *M.A.* **140** (1960), 351–9.
Ein metrischer Beitrag über Mahlersche S-Zahlen I, *J.M.* **203** (1960), 154–6.

VOLKMANN, B. The real cubic case of Mahler's conjecture, *Mathematika*, **8** (1961), 55–7.

Zur metrischen Theorie der *S*-Zahlen I, II, *J.M.* **209** (1962), 201–10; **213** (1964), 58–65.

WALDSCHMIDT, M. Solution d'un problème de Schneider sur les nombres transcendants, *C.R.* **271** (1970), 697–700.

Amélioration d'un théorème de Lang sur l'indépendance algébrique d'exponentielles, *C.R.* **272** (1971), 413–5.

Indépendance algébrique des valeurs de la fonction exponentielle, *Bull. Soc. Math. France*, **99** (1971), 285–304.

Propriétés arithmétiques des valeurs de fonctions méromorphes algébriquement indépendantes, *Acta Arith.* **23** (1973), 19–88.

Solution du huitième probleme de Schneider, *J. Number Th.* **5** (1973), 191–202.

WALLISSER, R. Zur Approximation algebraischer Zahlen durch arithmetisch charakterisierte algebraische Zahlen, *Arch. Math.* (*Basel*), **20** (1969), 384–91.

Zur Transzendenz der Werte der Exponentialfunktion, *Monatsh. Math.* **73** (1969), 449–60.

Über Produkte transzendenter Zahlen, *J.M.* **258** (1973), 62–78.

WEIERSTRASS, K. Zu Lindemann's Abhandlung 'Über die Ludolph'sche Zahl', *S.-B. Preuss Akad. Wiss.* (1885), 1067–85; = *Werke* II, 341–62.

WIRSING, E. Approximation mit algebraischen Zahlen beschränkten Grades, *J.M.* **206** (1961), 67–77.

On approximations of algebraic numbers by algebraic numbers of bounded degree, *Proc. Symposia Pure Math.* vol. 20 (Amer. Math. Soc. 1971), pp. 213–47.

FURTHER PUBLICATIONS

We give below a selected list of the new works on transcendental number theory that have appeared since the first printing; the list contains also a few older papers that were omitted originally. Abbreviations and other conventions have been adopted as before (see page 130).

ANDERSON, M. Inhomogeneous linear forms in algebraic points of an elliptic function, *Transcendence theory: advances and applications* (Academic Press, London, 1977), pp. 121–43.

Linear forms in algebraic points of an elliptic function, Ph.D. dissertation (Nottingham, 1978).

BAKER, A. Recent advances in transcendence theory, *Trudy Mat. Inst. Steklov*, **132** (1973), 67–9.

Some aspects of transcendence theory, *Astérisque* (*Soc. Math. France*), **24–5** (1975), 169–75.

The theory of linear forms in logarithms, *Transcendence theory: advances and applications* (Academic Press, London, 1977), pp. 1–27.

Transcendence theory and its applications, *J. Australian Math. Soc.* **25** (1978), 438–44.

BAKER, A. and COATES, J. Fractional parts of powers of rationals, *Math. P.C.P.S.*, **77** (1975), 269–79.

BAKER, A. and STEWART, C. L. Further aspects of transcendence theory, *Astérisque* (*Soc. Math. France*), **41–2** (1977), 153–63.

BAKER, R. C. On approximation with algebraic numbers of bounded degree, *Mathematika*, **23** (1976), 18–31.

Sprindžuk's theorem and Hausdorff dimension, *ibid.* 184–97.

Singular n-tuples and Hausdorff dimension, *Math. P.C.P.S.*, **81** (1977), 377–85.

Dirichlet's theorem on Diophantine approximation, *ibid.*, **83** (1978), 37–59.

BERTRAND, D. Equations différentielles algébriques et nombres transcendants dans les domaines complexe et p-adique, Thesis (Paris VI, 1975); and related work *C.R.* **279** (1974), 355–7; **282** (1976), 1399–401.

Séries d'Eisenstein et transcendance, *Bull. Soc. Math. France*, **104** (1976), 309–21.

Sous-groupes à un paramètre p-adique de variétés de groupe, *Invent. Math.* **40** (1977), 171–93.

Sur les dénominateurs des points rationnels des courbes elliptiques, *Astérisque* (*Soc. Math. France*), **41–2** (1977), 173–8.

Algebraic values of p-adic elliptic functions, *Transcendence theory: advances and applications* (Academic Press, London, 1977), pp. 149–58.

A transcendence criterion for meromorphic functions, *ibid.* pp. 187–93.

Approximations Diophantiennes p-adiques sur les courbes elliptiques admettant une multiplication complexe, *Compositio Math.* **37** (1978), 21–50.

BERTRAND, D. and FLICKER, Y. Z. Linear forms on Abelian varieties over local fields, *Acta Arith.* **38** (1980), 47–61.

BEUKERS, F. On the generalized Ramanujan–Nagell equation I, II, *Acta Arith.* **38** (1981), 389–410; **39** (1981), 113–23.

A note on the irrationality of $\zeta(2)$ and $\zeta(3)$, *Bull. London Math. Soc.* **11** (1979), 268–72.

Fractional parts of powers of rationals, *Math. P.C.P.S.* **90** (1981), 13–20.

BIJLSMA, A. On the simultaneous approximation of a, b and a^b, *Compositio Math.* **35** (1977), 99–111.

BROWNAWELL, W. D. Gelfond's method for algebraic independence, *Trans. Amer. Math. Soc.* **210** (1975), 1–26.

A measure of linear independence for some exponential functions, *Transcendence theory: advances and applications* (Academic Press, London, 1977), pp. 161–8.

Some remarks on semi-resultants, *ibid.* pp. 205–10.

On the Gelfond–Feldman measure of algebraic independence, *Compositio Math.* **38** (1979), 355–68.

BROWNAWELL, W. D. and KUBOTA, K. K. The algebraic independence of Weierstrass functions and some related numbers, *Acta Arith.* **33** (1977), 111–49.

BROWNAWELL, W. D., and WALDSCHMIDT, M. The algebraic independence of certain numbers to algebraic powers, *Acta Arith.* **32** (1977), 63–71.

BUNDSCHUH, P. Irrationalität und Transzendenz gewisser Riehen, *Math. Scand.* **28** (1971), 226–32.

Zu einem Transzendenzsatz der Herren Baron und Braune, *J.M.* **263** (1973), 183–8.

Zum Franklin-Schneiderschen Satz, *J.M.* **260** (1973), 108–18.

Zur Approximation gewisser p-adischer algebraischer Zahlen durch rationale Zahlen, *J.M.* **265** (1974), 154–9.

Zur simultanen Approximation von $\beta_0, \ldots, \beta_{n-1}$ und $\prod_{\nu=0}^{n-1} a_\nu^{\beta_\nu}$ durch algebraische Zahlen, *J.M.* **279** (1975), 99–117.

Transzendenzmasse in Körpern formaler Laurentreihen, *J.M.* **300** (1978), 411–32.

BUNDSCHUH, P. and WALLISSER, R. Algebraische Unabhängigkeit, P-adischer Zahlen, *M.A.* **221** (1976), 243–9.

CHUDNOVSKY, G. V. A measure of mutual transcendence for certain number classes (in Russian), *D.A.N.* **18** (1974), 771–4.

Baker's method in the theory of transcendental numbers (in Russian), *U.M.* **31** (1976), 281–2.

The Gelfond–Baker method in problems of Diophantine approximation, *Colloq. Math. Soc. János Bolyai*, vol. 13 (1976), pp. 19–30.

A new method for the investigation of arithmetical properties of analytic functions, *Ann. Math.* **109** (1979), 353–76.

CIJSOUW, P. L. Transcendence measures of exponentials and logarithms of algebraic numbers, *Compositio Math.* **28** (1974), 163–78.

Transcendence measures of certain numbers whose transcendency was proved by A. Baker, *ibid.* 179–94.

On the approximability of the logarithms of algebraic numbers, *Sém. Delange-Pisot-Poitou* **16** (1975), No. 19.

On the simultaneous approximation of certain numbers, *Duke Math. J.* **42** (1975), 249–57.

A transcendence measure for π, *Transcendence theory: advances and applications* (Academic Press, London, 1977), pp. 93–100.

Cijsouw, P. L. and Tijdeman, R. An auxiliary result in the theory of transcendental numbers II, *Duke Math. J.* **42** (1975), 239–47.

Cijsouw, P. L. and Waldschmidt, M. Linear forms and simultaneous approximations, *Compositio Math.* **34** (1977), 173–97.

Coates, J. and Lang, S. Diophantine approximation on Abelian varieties with complex multiplications, *Invent. Math.* **34** (1976), 129–33.

Cusick, T, W. Effective lower bounds for some linear forms, *Trans. Amer. Math. Soc.* **222** (1976), 289–301.

Davis, C. S. Rational approximations to e, *J. Australian Math. Soc.* **25** (1978), 497–502.

Davison, J. L. A series and its associated continued fraction, *Proc. Amer. Math. Soc.* **63** (1977), 29–32; and related work with W. W. Adams, *ibid.* **65** (1977), 194–8.

Demjanenko, V. A. Tate height and the representation of numbers by binary forms (in Russian), *I.A.N.* **38** (1974), 459–70.

Dubois, E. Application de la méthode de W. M. Schmidt à l'approximation de nombres algébriques dans un corps de fonctions de caractéristique zéro, *C.R.* **284** (1977), 1527–30.

Dubois, E. and Rhin, G. Approximations rationnelles simultanées de nombres algébriques réels et de nombres algébriques p-adiques, *Astérisque* (*Soc. Math. France*), **24–5** (1975), 211–27.

Sur la majoration de formes linéaires à coefficients algébriques réels et p-adiques. Demonstration d'une conjecture de K. Mahler, *C.R.* **282** (1976), 1211–14.

Durand, A. Quatre problèmes de Mahler sur la fonction ordre d'un nombre transcendant, *Bull. Soc. Math. France*, **102** (1974), 365–77; and related work *C.R.* **280** (1975), 309–11, 1085–8.

Note on rational approximations of the exponential function at rational points, *Bull. Australian Math. Soc.* **14** (1976), 449–55.

Indépendance algébrique de nombres complexes et critère de transcendance, *Compositio Math.* **35** (1977), 259–67.

Elianu, Jean. Sur le nombre $[a, a + \beta, a + 2\beta, \ldots, a + n\beta, \ldots]$. *Rev. Roumaine Math. Pures Appl.* **20** (1975), 1061–71.

Erdös, P. and Shorey, T. N. On the greatest prime factor of $2^p - 1$ for a prime p, and other expressions, *Acta Arith.* **30** (1976), 257–65.

Feldman, N. I. The periods of elliptic functions (in Russian), *Acta Arith.* **24** (1974), 477–89.

Estimates of linear forms in the logarithms of algebraic numbers, and some applications (in Russian), *Current problems in analytic number theory* (Minsk, 1974), pp. 244–68.

Rational approximations of algebraic numbers, *Colloq. Math. Soc. János Bolyai*, vol. 13 (1976), pp. 31–9.

Approximation of number π by algebraic numbers from special fields, *J. Number Theory*, **9** (1977), 48–60.

Flicker, Y. Z. On p-adic G-functions, *J. London Math. Soc.* **15** (1977), 395–402.

Transcendence theory over local fields, Ph.D. dissertation (Cambridge, 1978).

Algebraic independence by a method of Mahler, *J. Australian Math. Soc.* **27** (1979), 173–88.

Franklin, R. The transcendence of linear forms in $\omega_1, \omega_2, \eta_1, \eta_2, 2\pi i, \log \gamma$, *Acta Arith.* **26** (1975), 197–206; but see also **31** (1976), 143–52.

GALOČKIN, A. I. Lower bounds of polynomials in the values of a certain class of analytic functions (in Russian), *Mat. Sb.* **95** (1974), 396–417.

The approximation of the parameters of certain hypergeometric functions (in Russian), *Mat. Zametki*, **17** (1975), 103–12; and related work, *ibid.* **18** (1975), 541–52; **20** (1976), 35–45; *V.M.*, no. 6 (1978), 25–32.

GERRITZEN, L. Ein p-adischer Beweis für die Irrationalität der Nullstellen von Besselfunktionen, *M.A.* **226** (1977), 253–5.

GROSSMAN, E. H. Units and discriminants of algebraic number fields, *Comm. Pure Appl. Math.* **27** (1974), 741–7.

GYÖRY, K. Sur l'irreductibilité d'une classe des polynômes I, II, *Publ. Math. Debrecen*, **18** (1971), 289–307; **19** (1972), 293–326.

On polynomials with integer coefficients and given discriminant I, II, III, IV, V, *Acta Arith.* **23** (1973), 419–26; *Publ. Math. Debrecen* **21** (1974), 125–44; **23** (1976), 141–65; **25** (1978), 155–67; *Acta Math. Hungar.* **32** (1978), 175–90.

Polynomials with given discriminant, *Colloq. Math. Soc. János Bolyai*, vol. 13 (1974), 65–78.

Sur une classe des corps de nombres algébriques et ses applications, *Publ. Math. Debrecen* **22** (1975), 151–75.

Représentation des nombres entiers par des formes binaires, *ibid.* **24** (1977), 363–75.

GYÖRY, K. and LOVÁSZ, L. Representation of integers by norm forms II, *Publ. Math. Debrecen*, **17** (1970), 173–81; Part I, by K. Györy, *ibid.* **16** (1969), 253–63.

GYÖRY, K. and PAPP, Z. Z. Effective estimates for the integer solutions of norm form and discriminant form equations, *Publ. Math. Debrecen* **25** (1978), 311–25; and similar titles to appear.

GYÖRY, K., TIJDEMAN, R. and VOORHOEVE, M. On the equation $1^k + 2^k + \ldots + x^k = y^z$, *Acta Arith.* (to appear).

HARASE, T. The transcendence of $e^{\alpha\omega+\beta(2\pi i)}$, *J. Fac. Sci. Univ. Tokyo* **21** (1974), 279–85.

On the linear form of transcendental numbers $a_1\omega + a_2\pi + a_3 \log a_4$, *ibid.* **23** (1976), 435–52.

KLEIMAN, H. On the Diophantine equation $f(x, y) = 0$, *J.M.* **287** (1976), 124–31.

KOTOV, S. V. The law of the iterated logarithm for binary forms with algebraic coefficients (in Russian), *Dokl. Akad. Nauk BSSR* **17** (1973), 591–4.

The greatest prime factor of a polynomial (in Russian), *Mat. Zametki*, **13** (1973), 515–22; = **13** (1973), 313–17.

The Thue–Mahler equation in relative fields (in Russian), *Acta Arith.* **27** (1975), 293–315.

Über die maximale Norm der Idealteiler des Polynoms $\alpha x^m + \beta y^n$ mit den algebraischen Koeffizienten, *ibid.* **31** (1976), 219–30.

KUBOTA, K. K. On the algebraic independence of holomorphic solutions of certain functional equations and their values, *M.A.* **227** (1977), 9–50.

On a transcendence problem of K. Mahler, *Canadian J. Math.* **29** (1977), 638–47.

Linear functional equations and algebraic independence, *Transcendence theory: advances and applications* (Academic Press, London, 1977), pp. 227–9.

LANG, S. Diophantine approximation on toruses, *Amer. J. Math.* **86** (1964), 521–33.

Division points of elliptic curves and abelian functions over number fields, *ibid.* **97** (1972), 124–32.

Higher dimensional Diophantine problems, *Bull. Amer. Math. Soc.* **80** (1974), 779–87.

Diophantine approximation on abelian varieties with complex multiplication, *Advances in Math.* **17** (1975), 281–336.

Elliptic curves. Diophantine analysis (Springer-Verlag, Berlin, Heidelberg, 1978).

LeVeque, W. J. On the equation $y^m = f(x)$, *Acta Arith.* **9** (1964), 209–19.

Loxton, J. H. and van der Poorten, A. J. On algebraic functions satisfying a class of functional equations, *Aequationes Math.* **14** (1976), 413–20; and related work *ibid.* **15** (1977), 114–15.

On the growth of recurrence sequences, *Math. P.C.P.S.* **81** (1977), 369–76.

Arithmetic properties of certain functions in several variables I, II, III, *J. Number Theory*, **9** (1977), 87–106; *J. Australian Math. Soc.* (to appear); *Bull. Australian Math. Soc.* **16** (1977), 15–47.

Transcendence and algebraic independence by a method of Mahler, *Transcendence theory: advances and applications* (Academic Press, London, 1977), pp. 211–26.

Mahler, K. Arithmetische Eigenschaften der Lösungen einer Klasse von Funktionalgleichungen, *M.A.* **101** (1929), 342–66.

Über das Verschwinden von Potenzreihen mehrerer Veränderlichen in speziellen Punktfolgen, *M.A.* **103** (1930), 573–87.

Arithmetische Eigenschaften einer Klasse transzendental-transzendenter Funktionen, *M.Z.* **32** (1930), 545–85.

Lectures on transcendental numbers, *Proc. Symposia Pure Math.*, vol. 20 (Amer. Math. Soc., 1971), pp. 248–74.

The classification of transcendental numbers, *Proc. Symposia Pure Math.*, vol. 24 (Amer. Math. Soc., 1973), pp. 175–9.

A p-adic analogue to a theorem of J. Popken, *J. Australian Math. Soc.* **16** (1973), 176–84.

On rational approximations of the exponential function at rational points, *Bull. Australian Math. Soc.* **10** (1974), 325–35.

On the transcendency of the solutions of a special class of functional equations, *ibid.* **13** (1975), 389–410; corrigendum **14** (1976), 477–8.

A necessary and sufficient condition for transcendency, *Math. Comput.* **29** (1975), 145–53.

On a paper by A. Baker on the approximation of rational powers of e, *Acta Arith.* **27** (1975), 61–87.

On a class of transcendental decimal fractions, *Comm. Pure Appl. Math.* **29** (1976), 717–25; and a similar title in *Number theory and algebra* (Academic Press, New York, 1977), pp. 209–14.

Lectures on transcendental numbers, Lecture Notes in Mathematics, vol. 546 (Springer-Verlag, Berlin, 1976).

Makarov, Yu. N. On the estimates of the measure of linear independence for the values of E-functions (in Russian), *V.M.*, no. 2 (1978), 3–12.

Mandelbrojt, S. Sur la décroissance des coefficients supposés rationnels d'une fonction entière ne s'annulant pas en e, *C.R.* **278** (1974), 921–4.

Masser, D. W. *Elliptic functions and transcendence*, Lecture Notes in Mathematics, vol. 437 (Springer-Verlag, Berlin, 1975).

Linear forms in algebraic points of Abelian functions I, II, III, *Math. P.C.P.S.* **77** (1975), 499–513; **79** (1976), 55–70; *Proc. London Math. Soc.* **33** (1976), 549–64.

On the periods of Abelian functions in two variables, *Mathematika*, **22** (1975), 97–107.

A note on a paper of Franklin, *Acta Arith.* **31** (1976), 143–52.

Division fields of elliptic functions, *Bull. London Math. Soc.* **9** (1977), 49–53.

The transcendence of certain quasi-periods associated with Abelian functions in two variables, *Compositio Math.* **35** (1977), 239–58.

The transcendence of definite integrals of algebraic functions, *Astérisque* (*Soc. Math. France*), **41–2** (1977), 231–8.

Some vector spaces associated with two elliptic functions, *Transcendence theory: advances and applications* (Academic Press, London, 1977), pp. 101–19.

A note on Abelian functions, *ibid.*, pp. 145–7.

Diophantine approximation and lattices with complex multiplication, *Invent. Math.* **45** (1978), 61–82.

MEYER, Y. Nombres transcendants et répartition modulo 1, *Bull. Soc. Math. France*, **25** (1971), 143–9.

MIGNOTTE, M. Une généralisation d'un théorème de Cugiani–Mahler, *Acta Arith.* **22** (1972), 57–67.

Demonstration probabiliste d'un lemme combinatoire pour l'approximation diophantienne des nombres algébriques, *Colloq. Math.* **30** (1974), 109–19.

Approximations rationelles de π et quelques autres nombres, *Bull. Soc. Math. France*, **37** (1974), 121–32.

Quelques problèmes d'effectivite en theorie des nombres, Thesis (Paris XIII, 1974).

Quelques remarques sur l'approximation rationnelle des nombres algébriques, *J.M.* **269** (1974), 341–7.

Approximation des nombres algébriques par certaines suites de rationnels, *Rend. Circ. Mat. Palermo*, **24** (1975), 87–124.

A note on linear recursive sequences, *J. Australian Math. Soc.* **20** (1975), 242–4.

MIGNOTTE, M. and WALDSCHMIDT, M. Approximation des valeurs de fonctions transcendantes, *Indag. Math.* **37** (1975), 213–23.

Approximation simultanée de valeurs de la fonction exponentielle, *Compositio Math.* **34** (1977), 127–39.

Linear forms in two logarithms and Schneider's method, *M.A.* **231** (1978), 241–67.

MORENO, C. J. The values of exponential polynomials at algebraic points I, II, *Trans. Amer. Math. Soc.* **186** (1973), 17–31; *Diophantine approximation and its applications* (Academic Press, New York, 1973), pp. 111–28.

NESTERENKO, JU. V. Estimates for the orders of the zeros of analytic functions of a certain class and their applications in the theory of transcendental numbers (in Russian), *D.A.N.* **205** (1972), 292–5; and a similar title *I.A.N.* **41** (1977), 253–84.

An order function for almost all numbers (in Russian), *Mat. Zametki*, **15** (1974), 405–14.

NURMAGOMEDOV, M. S. and CHIRSKY, V. G. On arithmetical properties of values of hypergeometric functions (in Russian), *V.M.* **28**, no. 2 (1973), 38–45; see also [V.G. Chirsky] no. 5 (1978), 3–8.

van der Poorten, A. J. Hermite interpolation and p-adic exponential polynomials, *J. Australian Math. Soc.* **21** (1976), 12–26.

On Baker's inequality for linear forms in logarithms, *Math. P.C.P.S.* **80** (1976), 233–48.

Effectively computable bounds for the solutions of certain Diophantine equations, *Acta Arith.* **33** (1977), 195–207.

Linear forms in logarithms in the p-adic case, *Transcendence theory: advances and applications* (Academic Press, London, 1977), pp. 29–57.

van der Poorten, A. J. and Loxton, J. H. Computing the effectively computable bound in Baker's inequality for linear forms in logarithms, *Bull. Australian Math. Soc.* **15** (1976), 33–57; and related work, **16** (1977), 83–98; corrigendum, **17** (1977), 151–5.

Ramachandra, K., Shorey, T. N. and Tijdeman, R. On Grimm's problem relating to factorisation of a block of consecutive integers I, II, *J.M.* **273** (1975), 109–24; **288** (1976), 192–201.

Reyssat, E. Mesures de transcendance de nombres lies aux fonctions elliptiques, Thesis (Paris VI, 1977).

Salihov, V. H. The algebraic independence of the values of E-functions that satisfy first order linear differential equations (in Russian), *Mat. Zametki*, **13** (1973), 29–40; and related work, *D.A.N.* **235** (1977), 30–3.

Schinzel, A. and Tijdeman, R. On the equation $y^m = f(x)$, *Acta Arith.* **31** (1976), 199–204.

Schlickewei, H. P. On the fractional parts of the sum of powers of rational numbers, *Mathematika*, **22** (1975), 154–5.

Die p-adische Verallgemeinerung des Satzes von Thue–Siegel–Roth–Schmidt, *J.M.* **288** (1976), 86–105.

Linearformen mit algebraischen koeffizienten, *Manuscripta Math.* **18** (1976), 147–85; and related work, *Acta Arith.* **31** (1976), 389–98; **33** (1977), 183–5.

On norm form equations, *J. Number Theory*, **9** (1977), 370–80; and related work, *ibid.* 381–92; *Astérisque* (*Soc. Math. France*), **41–2** (1977), 267–71.

Schmidt, W. M. Approximation to algebraic numbers, *Enseignement Math.* **17** (1971), 187–253.

Zur Methode von Stepanov, *Acta Arith.* **24** (1973), 347–67; and related work, *J. Number Theory*, **6** (1974), 448–80.

Simultaneous approximation to algebraic numbers by elements of a number field, *Monatsh. Math.* **79** (1975), 55–66.

Rational approximation to solutions of linear differential equations with algebraic coefficients, *Proc. Amer. Math. Soc.* **53** (1975), 285–9.

Equations over finite fields. An elementary approach, Lecture Notes in Mathematics, vol. 536 (Springer-Verlag, Berlin, 1976).

On Osgood's effective Thue theorem for algebraic functions. *Comm. Pure Appl. Math.* **29** (1976), 749–63; and related work *Acta Arith.* **32** (1977), 275–96; *J. Australian Math. Soc.* **25** (1978), 385–422.

Schneider, Th. Eine Bemerkung zu einem Satz von C. L. Siegel, *Comm. Pure Appl. Math.* **29** (1976), 775–82.

Shidlovsky, A. B. The arithmetic properties of the values of analytic functions (in Russian). *Trudy Mat. Inst. Steklov*, **132** (1973), 169–202.

Shorey, T. N. On the sum $\sum_{k=1}^{3} |2^{\pi^k} - a_k|$, a_k algebraic numbers, *J. Number Theory*, **6** (1974), 248–60.

On linear forms in the logarithms of algebraic numbers, *Acta Arith.* **30** (1976), 27–42.

SHOREY, T. N. and TIJDEMAN, R. New applications of Diophantine approximations to Diophantine equations, *Math. Scand.* **39** (1976), 5–18.

On the greatest prime factors of polynomials at integer points, *Compositio Math.* **33** (1976), 187–95.

SHOREY, T. N., VAN DER POORTEN, A. J., TIJDEMAN, R. and SCHINZEL, A. Applications of the Gelfond–Baker method to Diophantine equations, *Transcendence theory: advances and applications* (Academic Press, London, 1977), pp. 59–77.

ŠMELEV, A. A. On a method of A. O. Gelfond in the theory of transcendental numbers (in Russian), *Mat. Zametki*, **10** (1971), 415–26; see also **11** (1972), 635–44.

Simultaneous approximations of the values of the exponential function at nonalgebraic points (in Russian), *V.M.* **27**, no. 4 (1972), 25–33; see also no. 6 (1972), 5–14, and *Ukrain. Mat. Z.* **27** (1975), 555–63.

A criterion for the algebraic dependence of transcendental numbers (in Russian), *Mat. Zametki*, **16** (1974), 553–62.

SPRINDŽUK, V. G. Square-free divisors of polynomials and class numbers of algebraic number fields (in Russian), *Acta Arith.* **24** (1973), 143–9.

Algebraic number fields with large class number (in Russian), *I.A.N.* **38** (1974), 971–82; = **8** (1974), 967–78.

The metric theory of Diophantine approximations (in Russian), *Current problems of analytic number theory* (Minsk, 1974), pp. 178–88; see also *Trudy Mat. Inst. Steklov*, **132** (1973), 137–42.

An effective analysis of the Thue and Thue–Mahler equations (in Russian), *Current problems of analytic number theory* (Minsk, 1974), pp. 199–222.

Representation of numbers by the norm forms with two dominating variables, *J. Number Theory*, **6** (1974), 481–6.

A hyperelliptic Diophantine equation and class numbers (in Russian), *Acta Arith.* **30** (1976), 95–108.

SPRINDŽUK, V. G. and KOTOV, S. V. An effective analysis of the Thue–Mahler equation in relative fields (in Russian), *Dokl. Akad. Nauk. BSSR*, **17** (1973), 393–5.

The approximation of algebraic numbers by algebraic numbers of a given field (in Russian), *ibid.* **20** (1976), 581–4.

SRINIVASAN, S. On algebraic approximations to $2^{n^k}(k = 1, 2, 3, \ldots)$, *India J. Pure Appl. Math.* **5** (1974), 513–23.

STARK, H. M. Some effective cases of the Brauer–Siegel theorem, *Invent. Math.* **23** (1974), 135–52; see also *Bull. Amer. Math. Soc.* **81** (1975), 961–72.

On complex quadratic fields with class-number two, *Math. Comp.* **29** (1975) 289–302.

STEWART, C. L. The greatest prime factor of $a^n - b^n$, *Acta Arith.* **26** (1975), 427–33.

Divisor properties of arithmetical sequences, Ph.D. dissertation (Cambridge, 1976).

On divisors of Fermat, Fibonacci, Lucas and Lehmer numbers, *Proc. London Math. Soc.* **35** (1977), 425–47.

Primitive divisors of Lucas and Lehmer numbers, *Transcendence theory: advances and applications* (Academic Press, 1977), pp. 79–92.

A note on the Fermat equation, *Mathematika*, **24** (1977), 130–2.

Algebraic integers whose conjugates lie near the unit circle, *Bull. Soc. Math. France*, **106** (1978), 169–76.

STOLARSKY, K. B. *Algebraic numbers and Diophantine approximation* (Marcel Dekker, New York, 1974).

TIJDEMAN, R. Old and new in number theory, *Nieuw Arch. Wisk.* **20** (1972), 20–30.

On the maximal distance between integers composed of small primes, *Compositio Math.* **28** (1974), 159–62; for applications see joint work with H. G. Meijer, *ibid.* **29** (1974), 265–86.

Applications of the Gelfond-Baker method to rational number theory, *Colloq. Math. Soc. János Bolyai*, vol. 13 (1974), pp. 399–416.

Some applications of Baker's sharpened bounds to Diophantine equations, *Sém. Delange-Pisot-Poitou*, **16** (1975), No. 24.

Hilbert's seventh problem: On the Gelfond–Baker method and its applications, *Proc. Symposia Pure Math.*, vol. 28 (Amer. Math. Soc., 1976), pp. 241–68.

On the equation of Catalan, *Acta Arith.* **29** (1976), 197–209.

VÄÄNÄNEN, K. On a conjecture of Mahler concerning the algebraic independence of the values of some E-functions, *Ann. Acad. Sci. Fenn. Ser.* AI, no. **512** (1972); and similar titles, *ibid.* no. **536** (1973), and Math. **1** (1975), pp. 93–109, 183–94.

On lower estimates for linear forms involving certain transcendental numbers, *Bull. Australian Math. Soc.* **14** (1976), 161–79.

On lower bounds for polynomials in the values of E-functions, *Manuscripta Math.* **21** (1977), 173–80; see also *J.M.* **296** (1977), 205–11.

VOORHOEVE, M., GYÖRY, K. and TIJDEMAN, R. On the diophantine equation $1^k + 2^k + \ldots + x^k + R(x) = y^z$ *Acta Math.* **143** (1979), 1–8.

WALDSCHMIDT, M. Propriétés arithmétiques de fonctions méromorphes, *C.R.* **273** (1971), 554–7.

Dimension algébrique de sous-groupes analytiques de variétés abéliennes, *C.R.* **274** (1972), 1681–3.

Utilisation de la méthode de Baker dans des problèmes d'indépendance algébrique, *C.R.* **275** (1972), 1215–17.

Approximation par des nombres algébriques des zeros de séries entières à coefficients algébriques, *C.R.* **279** (1974), 793–6.

Initiation aux nombres transcendants, *Enseignement Math.* **20** (1974), 53–85.

A propos de la méthode de Baker, *Bull. Soc. Math. France*, **37** (1974), 181–92.

Indépendance algébrique par la méthode de G. V. Chudnovsky, *Sém. Delange-Pisot-Poitou*, **16** (1975), no. 8.

Images de points algébriques par un sous-groupe analytique d'une variété de groupe, *C.R.* **281** (1975), 855–8.

Dimension algébrique de sous-groupes analytiques de variétés de groupe, *Ann. Inst. Fourier, Grenoble*, **25** (1975), 23–33.

Some topics in transcendental number theory, *Colloq. Math. Soc. Janós Bolyai*, vol. 13 (1976), pp. 417–27.

Rapport sur la transcendance, *Astérisque* (*Soc. Math. France*), **41–2** (1977), 127–34.

On functions of several variables having algebraic Taylor coefficients, *Transcendence theory: advances and applications* (Academic Press, London, 1977), pp. 169–86; see also articles in *Lecture Notes in Mathematics*, vol. 524 (1976), vol. 578 (1977) (Springer-Verlag, Berlin).

Les travaux de G. V. Chudnovsky sur les nombres transcendants, *ibid.* vol. 567 (1977), pp. 274–92.

Transcendence measures for exponentials and logarithms, *J. Australian Math. Soc.* **25** (1978), 445–65.

Simultaneous approximations of numbers connected with the exponential function, *ibid.* 466–78.

A lower bound for linear forms in logarithms, *Acta Arith.* **37** (1980), 257-83.

WALLISSER, R. Über die arithmetische Natur der Werte der Lösungen einer Funktionalgleichung von H. Poincaré, *Acta Arith.* **25** (1974), 81–92.

WÜSTHOLZ, G. Zum Franklin-Schneiderschen Satz, *Manuscripta Math.* **20** (1977), 335–54.

Linearformen in Logarithmen von U-Zahlen mit ganzzahligen Koeffizienten, *J.M.* **300** (1978), 138–50.

NEW DEVELOPMENTS

We give a brief summary of some of the progress in transcendental number theory that has been made since this book was written. It has in fact been a very active field of research and only a few of the new results will be mentioned here. A full account of much of the ground that has been won recently can be found in *Transcendence theory: advances and applications* (Academic Press, London and New York, 1977), Proceedings of a conference held in Cambridge in 1976, edited by A. Baker and D. W. Masser; it will be referred to briefly as *TTAA*.

1. Linear forms in logarithms

The inequalities for Λ recorded after the enunciation of Theorem 3.1 have been much improved. It is shown in Chapter 1 of *TTAA* that if α_j and β_j are algebraic numbers with heights at most A_j (≥ 4) and B (≥ 4), and if the field generated by the α's and β's over the rationals has degree at most d, then either $\Lambda = 0$ or $|\Lambda| > (B\Omega)^{-C\Omega \log \Omega'}$, where $\Omega = \log A_1 \ldots \log A_n$, $\Omega' = \Omega/\log A_n$ and $C = (16nd)^{200n}$. Furthermore, in the rational case, that is when $\beta_0 = 0$ and $\beta_1, \ldots, \beta_n$ are rational integers, the bracketed factor Ω has been eliminated to yield $|\Lambda| > B^{-C\Omega \log \Omega'}$. The latter result is best possible with respect to A_n and B, and likewise with respect to $A_1, \ldots, A_{n-1}$ except for the second-order term $\log \Omega'$. The proof involves, in particular, an improved version, due to Tijdeman, of Lemma 1 on page 25, a new idea of Shorey on the size of the inductive steps which occur in Lemma 6 on page 31, and some rather deep developments concerning the argument on pages 34 and 35 involving more widespread use of Kummer theory. Refinements in the results, relating especially to the expression for C, have been announced by several writers.†

There is also an extensive p-adic theory of linear forms in logarithms. The subject was initiated by Mahler in the 1930s when he

† cf. *J. Australian Math. Soc.* **25** (1978), 445–78.

obtained p-adic analogues of both the Hermite–Lindemann and the Gelfond–Schneider theorems; in fact, in the course of the work, Mahler laid the foundations of the p-adic theory of analytic functions that has been fundamental to all later studies in the field. A good account of the subject is given by van der Poorten in Chapter 2 of *TTAA*; in particular he establishes there an estimate for the p-adic valuation of Λ of essentially the same degree of precision as that quoted earlier in the Archimedean case.

2. Diophantine equations

The sharpened estimates for Λ and, in particular, their best possible dependence on A_n, have led to some remarkable developments in connection with the theory of Diophantine equations.

First, they yield at once an explicit upper bound for all solutions in integers $x > 1$, $y > 1$, $n > 2$ of the equation $ax^n - by^n = c$, where $a, b, c\ (\neq 0)$ are any given integers. In fact, assuming, as we may without loss of generality, that a and b are positive, and that $y \geq x$, the equation gives $|\Lambda| \ll y^{-n}$, where $\Lambda = \log(a/b) + n\log(x/y)$, and the implied constant depends only on a, b and c. But, from the results recorded in §1, we have $\log|\Lambda| \gg -\log y \log n$; a comparison of estimates yields a bound for n in terms of a, b, c, and Theorem 4.2 then furnishes bounds for x and y.

A similar argument was employed by Tijdeman† to show that the famous Catalan equation $x^p - y^q = 1$ has only finitely many solutions in integers x, y, p, q (all > 1). Here we can assume that p, q are odd primes, and it will suffice to treat the case $p > q$. Then $x = kX^q + 1$, $y = lY^p - 1$ for some integers X, Y, where k is 1 or $1/p$ and l is 1 or $1/q$. Plainly we have

$$|p\log x - q\log y| \ll y^{-q},$$

and thus $|\Lambda| \ll (kX^q)^{-1}$, where

$$\Lambda = p\log k - q\log l + pq\log(X/Y).$$

Hence, by §1, we obtain $q \ll (\log p)^4$. Further, the linear form $\Lambda' = p\log(x/Y^q) - q\log l$ satisfies $|\Lambda'| \ll (lY^p)^{-1}$, and so again by §1, we have $p \ll q(\log p)^3$. Thus p and q are bounded, and so, by Theorem 4.2, also x and y are bounded, as required.

Many novel results have been obtained subsequent to Tijdeman's discovery; see especially Chapter 3 of *TTAA*. In particular, Schinzel

† *Acta Arith.* **29** (1976), 197–209.

and Tijdeman† have effectively resolved the equation $y^m = f(x)$, of hyperelliptic type, in integers $x, y > 1, m > 2$. Shorey and Tijdeman‡ have dealt with the equation $y^m = x^n + x^{n-1} + \ldots + 1$ in integers x, y, m, n (all > 1) when either x is fixed or $n + 1$ or y has a fixed prime factor. Győry, Tijdeman and Voorhoeve§ have shown that for any integer $k \geq 6$ the equation $y^m = 1^k + \ldots + x^k$ has only finitely many solutions in integers $x \geq 1, y \geq 1, m > 1$ all of which can, in principle, be effectively determined. Stewart|| has proved that the Fermat equation $x^n + y^n = z^n$ has only finitely many solutions in positive integers x, y, z and n (>2), provided that $|x - y|$ is bounded. van der Poorten¶ has used p-adic analysis to effectively resolve the equation $x^p - y^q = z^r$ in integers x, y, z, p, q (all > 1), where z is composed solely of powers of a fixed set of primes, and r is the least common multiple of p and q. And Győry and Papp†† have recently made valuable progress in connection with the problem of effectively solving equations of norm form in several variables of the kind referred to on page 68 of Chapter 7.

It should also be noted that the work of Schmidt discussed in Chapter 7 has been extensively generalized in the p-adic domain by Dubois and Rhin,‡‡ and by Schlickewei.§§

3. Elliptic and Abelian functions

Much research has been carried out in the last few years on the work of Chapter 6. In fact the results have been generalized in three directions, namely quantitatively, p-adically and in the context of Abelian varieties. Of particular interest are quantitative versions of Masser's result cited at the end of page 64; see the paper by Anderson in Chapter 7 of *TTAA*. The studies lead to delicate questions on the division value properties of the Weierstrass functions, and, in particular, some recent theorems of Bashmakov and Ribet on the subject have proved useful. The methods can be generalized, as Masser,|||| Coates and Lang¶¶ have shown, to deal with Abelian functions,

† *Acta Arith.* **31** (1976), 199–204.
‡ *Math. Scand.* **39** (1976), 5–18.
§ *Acta Arith.* (to appear).
|| *Mathematika* **24** (1977), 130–2. A similar result was obtained independently by Inkeri and van der Poorten.
¶ *Acta Arith.* **33** (1977), 195–207.
†† *Publ. Math. Debrecen* **25** (1978), 311–25.
‡‡ *Astérisque* (*Soc. Math. France*), **24–5** (1975), 211–27.
§§ *J.M.* **288** (1976), 86–105.
|||| *Invent. Math.* **45** (1978), 61–82.
¶¶ *Invent. Math.* **34** (1976), 129–33.

provided that one makes suitable assumptions concerning complex multiplications, and the theory has been carried over to the p-adic domain for a wide class of primes p by Bertrand and Flicker.†

Other noteworthy results are a p-adic version of Schneider's Theorem 6.2 obtained by Bertrand,‡ a solution to the open problem mentioned on page 57 in the case $p = 2$ by Masser,§ his extension of Theorem 6.6 to include also the number $2\pi i$,|| and a striking theorem of Chudnovsky¶ to the effect that if g_2 and g_3 are algebraic then the transcendence degree of the field generated by $\omega_1, \omega_2, \eta_1, \eta_2$ over the rationals is at least 2. As an immediate corollary one obtains the transcendence of $\Gamma(\frac{1}{4})$; this follows on considering the curve $y^2 = 4x^3 - 4x$ for which one can take $\omega_1 = (\Gamma(\frac{1}{4}))^2/\sqrt{(8\pi)}$, $\omega_2 = i\omega_1$, $\eta_1 = \pi/\omega_1$ and $\eta_2 = -i\eta_1$. Similarly from the curve $y^2 = 4x^3 - 4$ one deduces that $\Gamma(\frac{1}{3})$ is transcendental.

4. Arithmetical properties of meromorphic functions

An excellent account of the Siegel–Shidlovsky theorems is given by Mahler in his tract *Lectures on transcendental numbers* (Lecture Notes in Mathematics, vol. 546, Springer-Verlag, Berlin, 1976). The volume contains also some notable examples of entire transcendental functions which, together with their derivatives, assume algebraic values at all algebraic points, and also an interesting appendix giving an historical survey of the classical proofs of the transcendence of e and π. The papers of Waldschmidt and Bertrand in Chapters 11 and 12 of *TTAA* contain new results on meromorphic functions of finite order.

The problem of estimating the quantity $P(E_1, \ldots, E_n)$ referred to on page 117 has been studied by Nesterenko,†† and he has obtained bounds with an explicit dependence on the degree of P; further work in this connection, and also in relation to Chapter 10, has been carried out by Väänänen.‡‡ The question of estimating similar expressions for G-functions has been investigated by Nurmagomedov, Galočkin and, in the p-adic domain, by Flicker.§§ Incidentally, no

† *Acta Arith.* (to appear).

‡ Chapter 9 of *TTAA*, and *Invent. Math.* **40** (1977), 171–93.

§ *Mathematika*, **22** (1975), 97–107.

|| Chapter 6 of *TTAA*.

¶ See the article by Waldschmidt in *Lecture Notes in Mathematics*, vol. 567 (Springer-Verlag, Berlin, 1977), pp. 274–92.

†† *I.A.N.* **41** (1977), 253–84.

‡‡ *Manuscripta Math.* **21** (1977), 173–80; *Bull. Australian Math. Soc.* **14** (1976), 161–79.

§§ *J. London Math. Soc.* **15** (1977), 395–402.

satisfactory p-adic generalization of the Siegel–Shidlovsky theory has been given as yet.

In another direction, an old and previously neglected method of Mahler on functions satisfying certain functional equations has become the focus of much activity in the last few years. The principles underlying the method are simpler than those of the Siegel–Shidlovsky theory, and they have proved to be capable of considerable generalization; see the survey articles of Loxton and van der Poorten and of Kubota in Chapters 15 and 16 of *TTAA*. Typical examples of the results furnished by the method are the transcendence of the Fredholm series $\sum z^{2^n}$ for all algebraic z with $0 < |z| < 1$, and of $\sum [n\omega]z^n$ for all such z and any irrational ω. It also yields several new instances of algebraically independent numbers.

5. Further works

Stewart† has utilized the estimates for linear forms in logarithms quoted earlier to give much new information on divisor properties of arithmetical sequences, and, in particular, on Lucas and Lehmer numbers.

Chudnovsky has announced some extensive developments of the method of Gelfond discussed in Chapter 12 leading to the algebraic independence of three or more numbers from certain sets, rather than just two as occur in Theorems 12.1 and 12.2.‡

R. C. Baker§ has obtained new results on T-numbers, and also on a conjecture of Schmidt and the author concerning the metrical theory discussed in Chapter 10.

Stepanov, and subsequently Bombieri and Schmidt, used transcendence techniques to furnish a new proof of the Riemann hypothesis for curves which is of a much more elementary nature than that given originally.||

And finally we mention that Apéry has just established the irrationality of $\zeta(3)$.

† *Acta Arith.* **26** (1975), 427–33; *Proc. London Math. Soc.* **35** (1977), 425–47; see also Chapter 4 of *TTAA*.

‡ See again the article by Waldschmidt in *Lecture Notes in Mathematics*, vol. 567 (Springer-Verlag, Berlin, 1977), pp. 274–92.

§ *Mathematika*, **23** (1976), 18–31, 184–97.

|| See Schmidt's tract, vol. 536 (1976) of *Lecture Notes in Mathematics*. For an especially simple proof in the case of elliptic curves see *Astérisque* (*Soc. Math. France*), **24–5** (1975), 173–5.

6. Additional remarks (1990)

A natural sequel to *TTAA* (see page 155) is *New advances in transcendence theory* (Cambridge University Press, 1988), Proceedings of a Symposium held in Durham in 1986, edited by A. Baker; we shall refer to the volume briefly as *NATT*. It covers the most significant discoveries of recent years.

The papers in *NATT* by Wüstholz and by Philippon and Waldschmidt establish, independently, that the second-order term $\log \Omega'$ can be eliminated from the estimates for $|\Lambda|$ discussed on page 155. The arguments depend on the theory of multiplicity estimates on group varieties. This is an important new instrument in transcendence originating from techniques involving commutative algebra introduced by Nesterenko;† it has been developed by Brownawell and Masser,‡ Masser and Wüstholz,§ Philippon|| and, most notably, by Wüstholz.¶ The work has led to a remarkable synthesis of many aspects of the subject in terms of algebraic groups.

There are several substantial memoirs in *NATT* relating to Diophantine analysis; they include, in particular, an article by Evertse, Györy, Stewart and Tijdeman surveying the fertile field of *S*-unit equations and a paper by Odoni furnishing an application to a question of Serre on modular forms. It is apparent, *vide* the tract†† *Exponential Diophantine equations* (Cambridge University Press, 1986) by T. N. Shorey and R. Tijdeman, that the topic has become an unusually rich area of research during the past decade. In another exciting development, two new proofs of the Mordell conjecture on the finiteness of the number of rational points on algebraic curves of genus >1, first established by Faltings in 1983, have recently been obtained by Masser and Wüstholz and by Vojta using transcendence methods.

The study of *E*-functions and *G*-functions has continued to be a source of much activity. Again, a good account is provided by *NATT*; there are, especially, some beautiful results of Beukers and Wolfart specifying precisely when the classical hypergeometric functions can

† *I.A.M.* **41** (1977), 253-84.
‡ *J.M.* **314** (1980), 200-16; *Duke Math. J.* **47** (1980), 273-95.
§ *Invent. Math.* **64** (1981), 489-516; **80** (1985), 233-67.
|| *Bull. Soc. Math. France* **114** (1986), 355-83.
¶ *J.M.* **317** (1980), 102-19; *Ann. Math.* **129** (1989), 471-517.
†† Here, much rests on the *p*-adic estimate of van der Poorten mentioned on page 156. Yu Kunrui has recently given an improved version (*Compositio Math.*, to appear) overcoming some errors in van der Poorten's demonstration; see the discussion in *NATT*.

take algebraic values. A comprehensive exposition of the basic Siegel-Shidlovsky theory is given in *Transcendental numbers* (de Gruyter, 1989), Studies in mathematics 12, by A. B. Shidlovsky.

The metrical conjecture concerning $(\psi(h))^n$ mentioned on page 95 has been proved by V. I. Bernik; references to this and also to his solution to a problem of Baker and Schmidt on Hausdorff dimension are given in his paper in *NATT*.

The famous problem of Gauss of determining effectively all the imaginary quadratic fields with a given class number, earlier resolved for class numbers 1 and 2 as described in Chapter 5, has now been resolved in general through the theory of elliptic curves, by Goldfeld, Gross and Zagier.‡‡

‡‡ *C.R.* **297** (1983), 85-7; *Bull. Amer. Math. Soc.* **13** (1985), 23-37.

Some developments since 1990

David Masser

We comment on each chapter in turn, taking into account also the New Developments ND including the Additional Remarks (1990) AR. We refer only to books that have appeared since 1990.

In Chapter 1 the Lindemann–Weierstrass Theorem 1.4 is equivalent to the algebraic independence of $e^{\alpha_1}, \ldots, e^{\alpha_n}$ for any algebraic $\alpha_1, \ldots, \alpha_n$ linearly independent over $\mathbf{Q}$. An elliptic analogue of this was proved by Philippon and Wüstholz independently in 1983. Namely suppose $\wp(z)$ is a Weierstrass function with algebraic invariants and complex multiplication by an imaginary quadratic field k. Then $\wp(\alpha_1), \ldots, \wp(\alpha_n)$ are defined and algebraically independent for any algebraic $\alpha_1, \ldots, \alpha_n$ linearly independent over k. For more see [BW].

Concerning Chapter 2, see [Wa] for an encyclopaedic account of developments around the exponential function.

There are also elliptic analogues of Theorem 2.1. In the case of complex multiplication CM this is already mentioned in Chapter 6; but also when there is no CM (in which case one could write $k = \mathbf{Q}$). Thus if $u_1, \ldots, u_n$ are complex numbers, linearly independent over k, such that $\wp(u_1), \ldots, \wp(u_n)$ are algebraic, then $1, u_1, \ldots, u_n$ are linearly independent over the field $\overline{\mathbf{Q}}$ of all algebraic numbers. For this as well as abelian analogues and more, see the comments below on Chapter 6.

For an account of the more recent developments in Chapter 3 see also [BW].

There has been much work on the Diophantine equations mentioned in Chapter 4, for which one can consult [EG]. But perhaps the most unexpected is Mihăilescu's proof that the Catalan equation $p^s - q^r = 1$ has no solutions except that arising from $9 - 8 = 1$. For an excellent account see [BBM].

Regarding Chapter 5, the work of Goldfeld, Gross and Zagier was already mentioned in the AR, and all imaginary quadratic fields with class number up to 100 have been listed; see also [BW] for more details.

The elliptic analogues of Theorem 2.1 referred to in Chapter 6 and mentioned above, as well as the abelian analogues and more, are all special cases of the Analytic Subgroup Theorem of Wüstholz, valid for any complex commutative algebraic group G. If it has dimension N, there is an exponential map $\exp = \exp_G$ from $\mathbf{C}^N$ to the complex points of G, and if G is defined over $\overline{\mathbf{Q}}$ then this map can also be normalized over $\overline{\mathbf{Q}}$. One takes u in $\mathbf{C}^N$ with $g = \exp(u)$ a non-zero algebraic point of G. The result says that if g lies in some analytic subgroup Z of G, also defined over $\overline{\mathbf{Q}}$, then Z must contain a non-zero connected algebraic subgroup of G also defined over $\overline{\mathbf{Q}}$.

Theorem 2.1 is the special case $G = \mathbf{G}_{\mathrm{a}} \times \mathbf{G}_{\mathrm{m}}^n$, where (over the complex numbers) $\mathbf{G}_{\mathrm{a}}$ is the additive group $\mathbf{C}$ and $\mathbf{G}_{\mathrm{m}}$ is the multiplicative group $\mathbf{C}^*$, with $\exp(z_0, z_1, \ldots, z_n) = (z_0, e^{z_1}, \ldots, e^{z_n})$.

See [BW] for a detailed account, as well as applications such as the transcendence of integrals of algebraic functions over paths; also for new kinds of applications to estimates for isogenies, endomorphisms, factorizations, and polarizations of abelian varieties and also the Tate Conjecture. More recently see [HW] for more about integrals in the language of motives.

See also [T] for yet further applications.

The decisive developments concerning Chapter 7 are around Schmidt's Subspace Theorem which generalizes Theorems 7.1 and 7.2. Namely let $L_1, \ldots, L_n$ be linear forms in $n \geq 2$ variables with coefficients in $\overline{\mathbf{Q}}$ and linearly independent over $\overline{\mathbf{Q}}$. Then for any $\delta > 0$, the non-zero $\mathbf{x} = (x_1, \ldots, x_n)$ in $\mathbf{Z}^n$ with

$$|L_1(\mathbf{x}) \cdots L_n(\mathbf{x})| < (\max\{|x_1|, \ldots, |x_n|\})^{-\delta}$$

lie in a finite union of subspaces $X \neq \mathbf{Q}^n$ of $\mathbf{Q}^n$.

There are also important p-adic versions by Schlickewei and others. There is a good account in [BG] together with the applications to Diophantine equations of norm form.

See also [CZ] for some remarkable applications to integer points on algebraic surfaces.

For a recent exposition of some of the topics of Chapter 8 see [B].

Similarly regarding Chapter 9 see [H].

The Littlewood Conjecture mentioned in Chapter 10 has attracted much attention from the viewpoint of Ergodic Theory, as explained in [EW].

In connexion with Chapter 11 on E-functions, the parallel work on G-functions is already mentioned in ND. For an account of the geometric side of the latter theory one can consult [A]; this also contains the first appearance of the André–Oort Conjecture.

Also in the ND, Mahler's Method is mentioned in connexion with its relationship to E-functions. The interested reader can consult [N].

As remarked in AR, the topic of algebraic independence in Chapter 12 has been greatly developed, and one can find a useful account of this in [NP].

Regarding the work of Ax mentioned in Chapter 12, this has been greatly generalized in connexion with the Pila–Zannier strategy for 'unlikely intersections'. There is an excellent account in [Z] (including the Zilber–Pink Conjectures) – see also [JW] (focusing more on André–Oort).

In ND the transcendence of $\Gamma(1/3)$ and $\Gamma(1/4)$ and the irrationality of $\zeta(3)$ are mentioned. At the time of writing we still do not know about $\Gamma(1/5)$ and $\zeta(5)$.

Finally one should not forget [Wü], a broad collection of articles written to celebrate Baker's sixtieth birthday. And see [M] for an account of his life and works.

Additional Bibliography

[A] Andre, Y. *G-functions and geometry*, Aspects of mathematics, vol. E13 (Friedr. Vieweg & Sohn, 1989).

[BW] Baker, A. and Wusthölz, G. *Logarithmic forms and diophantine geometry*, New Mathematical Monographs, vol. 9 (Cambridge University Press, 2007).

[BBM] Bilu, Y. F., Bugeaud, Y. and Mignotte, M. *The problem of Catalan* (Springer, 2014).

[BG] Bombieri, E. and Gubler, W. *Heights in diophantine geometry*, New Mathematical Monographs, vol. 4 (Cambridge University Press, 2006).

[B] Bugeaud, Y. *Approximation by algebraic numbers*, Cambridge Tracts in Mathematics, vol. 160 (Cambridge University Press, 2004).

[CZ] Corvaja, P. and Zannier, U. *Applications of diophantine approximation to integral points and transcendence*, Cambridge Tracts in Mathematics, vol. 212 (Cambridge University Press, 2018).

[EG] Evertse, J.-H. and Győry, K., *Unit equations in diophantine number theory*, Cambridge Studies in Advanced Mathematics, vol. 146 (Cambridge University Press, 2015).

[EW] Einsiedler, M. and Ward, T. *Ergodic theory: with a view towards number theory*, Graduate Texts in Mathematics, vol. 259 (Springer, 2011).

[H] Harman, G. *Metric number theory*, LMS Monographs, vol. 18 (Oxford University Press, 1998).

[HW] Huber, A. and Wüstholz, G. *Transcendence and linear relations of 1-periods*, Cambridge Tracts in Mathematics, vol. 227 (Cambridge University Press, 2022).

[JW] Jones, G. O. and Wilkie, A. (eds). *O-minimality and diophantine geometry*, LMS Lecture Notes, vol. 421 (Cambridge University Press, 2015).

[M] Masser, D. Alan Baker FRS, 1939–2018, *Bull. London Math. Soc.* (Dec. 2021), available at https://doi.org/10.1112/blms.12553

[NP] Nesterenko, Y. V. and Philippon, P. (eds). *Introduction to algebraic independence theory*, Lecture Notes in Math., vol. 1752 (Springer, 2001).

[N] Nishioka, K. *Mahler functions and transcendence*, Lecture Notes in Math., vol. 1631 (Springer, 1996).

[T] Tretkoff, P. *Periods and special functions in transcendence* (World Scientific, 2017).

[Wa] Waldschmidt, M. *Diophantine approximation on linear algebraic groups*, Grundlehren der mathematischen Wissenschaften, vol. 326 (Springer, 2000).

[Wü] Wüstholz, G. *A panorama of number theory, or the view from Baker's garden* (Cambridge University Press, 2002).

[Z] Zannier, U. *Some problems of unlikely intersections in arithmetic and geometry*, Annals of Math. Studies, vol. 181 (Princeton University Press, 2012).

INDEX

Entries for authors refer to places where either their name or one of their publications has been cited.

Printed in the United States
by Baker & Taylor Publisher Services